SOLVING PROBLEMS IN MATHEMATICS

SERIES EDITOR : JAMES FULTON
University of Edinburgh

8

Y0-CCO-249

SOLVING PROBLEMS IN
ADVANCED CALCULUS: I

TITLES IN THIS SERIES

Solving Problems in

ADVANCED CALCULUS: I

R. P. GILLESPIE
formerly University of Glasgow

OLIVER & BOYD
EDINBURGH

OLIVER AND BOYD
Tweeddale Court
14 High Street
Edinburgh EH1 1YL

A Division of Longman Group Limited

First published 1972

ISBN 0 05 002430 2

Printed in Great Britain by R. & R. Clark Ltd, Edinburgh

PREFACE

This book is based on lectures which I have given at the University of Glasgow to second and third year students, both at the Honours and at the Ordinary Degree level. It is hoped that the book will be of use not only to students of Pure Mathematics but also to students of Applied Mathematics, Engineering and Statistics.

Reference is made in the latter half of the book to the volume *Vector Algebra*, by E. M. Patterson, in the same series. For background theory, the student is recommended to read the relevant books in the *University Mathematical Texts* series, also published by Oliver and Boyd.

I have to thank the University Court of the University of Glasgow for permission to use many problems taken from their degree papers. I should also like to thank Dr James Fulton, the general editor of the series for his many helpful suggestions and my son, Dr T. A. Gillespie, who has read the proofs and checked the answers to the problems.

R. P. GILLESPIE

CONTENTS

Chapter I

PARTIAL DIFFERENTIATION

1. Functions of Several Variables: Surfaces

Let A and B be two sets of objects and suppose that there is a *relation* f between the objects of the set A and the objects of the set B, such that to any object of A there corresponds a unique object of B. Then we say that the relation f is a *function* with *domain* A and *range* B. In this chapter the members of the set A will consist of ordered sets of real numbers $(x_1, x_2, \ldots, x_n)$ and the members of B of real numbers z. We write $z = f(x_1, x_2, \ldots, x_n)$. We frequently call $x_1, \ldots, x_n$, z *variables*; here $x_1, \ldots, x_n$ are called *independent* variables while z is the *dependent* variable. We begin by discussing functions of two independent variables, that is, functions of the form $z = f(x, y)$. The type of function with which we shall be most concerned is that where the relation between the ordered pair (x, y) and z is given by means of a formula; e.g., $z = x^2 - y^2$.

If $z = f(x, y)$, the set of points whose Cartesian coordinates in three-dimensional space R^3 are (x, y, z) is called the *surface* associated with the function f and we say that the relation $z = f(x, y)$ is the *equation* of the surface. The function may be defined *implicitly* by means of an equation $F(x, y, z) = 0$; e.g., $F = x^2 + y^2 + z^2 - a^2 = 0$ defines the *two-valued* function $z = \pm\sqrt{(a^2 - x^2 - y^2)}$, and the function is represented by the surface with equation $F(x, y, z) = 0$.

In this paragraph we recall some of the surfaces which will be required later. We mention first two alternative coordinate systems in three-dimensional space.

M is the projection of the point P, with Cartesian coordinates (x, y, z), in the plane XOY. If $OM = \rho$, $\angle XOM = \phi$, then (ρ, ϕ, z) are *cylindrical coordinates* of P, and projection of OM on OX, OY gives $x = \rho \cos \phi$, $y = \rho \sin \phi$. If $OP = r$, $\angle ZOP = \theta$ and, as before, $\phi = \angle XOM$, (r, θ, ϕ) are *spherical polar coordinates* of P. It follows at once from Fig. 1 (p. 2) that $x = r \sin \theta \cos \phi$, $y = r \sin \theta \sin \phi$, $z = r \cos \theta$.

1

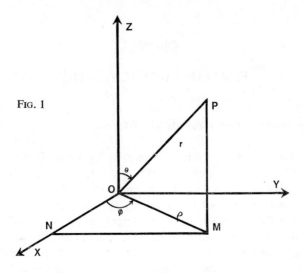

FIG. 1

(i) The surface $ax+by+cz+d = 0$ is a *plane*, whose normal has direction cosines proportional to (a, b, c).

(ii) The surface $(x-a)^2+(y-b)^2+(z-c)^2 = r^2$ is a *sphere* with centre the point (a, b, c) and radius of length r.

(iii) The surface $f(x, y) = 0$ is a *cylinder* with generators parallel to the z-axis through the curve $f(x, y) = 0$ in the plane $z = 0$.

(iv) If $F(x, y, z)$ is homogeneous of degree n in x, y, z, i.e. if

$$F(\lambda x, \lambda y, \lambda z) \equiv \lambda^n F(x, y, z),$$

for some value of n, the surface $F(x, y, z) = 0$ is a *cone* with vertex at the origin.

(v) The surface $F[\sqrt{(x^2+y^2)}, z] = 0$ is a *surface of revolution* about the z-axis.

(vi) If $F(x, y, z)$ is a polynomial of degree two in x, y and z, the surface with equation $F(x, y, z) = 0$ is called a *quadric* surface. Quadrics are of the following types.

(a) *Ellipsoid*

The equation

$$\frac{x^2}{a^2}+\frac{y^2}{b^2}+\frac{z^2}{c^2} = 1$$

represents an ellipsoid (if $a = b = c$, we have a sphere). Any plane section of an ellipsoid is an ellipse (or in some cases a circle).

The origin is the centre of this ellipsoid. (Fig. 2, below.)

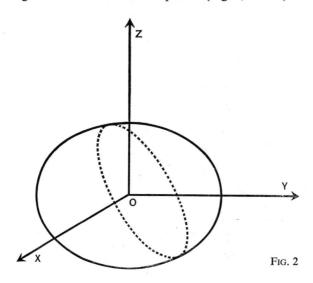

Fig. 2

(b) Hyperboloid

We can distinguish between two types of hyperboloid. The first type, a *hyperboloid of one sheet,* is exemplified by the surface with equation

$$\frac{x^2}{a^2} + \frac{y^2}{b^2} - \frac{z^2}{c^2} = 1.$$

Planes parallel to $z = 0$ intersect the surface in ellipses, while planes parallel to $x = 0$ and $y = 0$ intersect the surface in hyperbolas. (Fig. 3, p. 4.)

The second type, a *hyperboloid of two sheets,* is exemplified by the surface with equation

$$-\frac{x^2}{a^2} - \frac{y^2}{b^2} + \frac{z^2}{c^2} = 1.$$

Again, sections parallel to $z = 0$ are ellipses, while sections parallel to the planes $y = 0$ and $x = 0$ are hyperbolas. (Fig. 4, p. 5.)

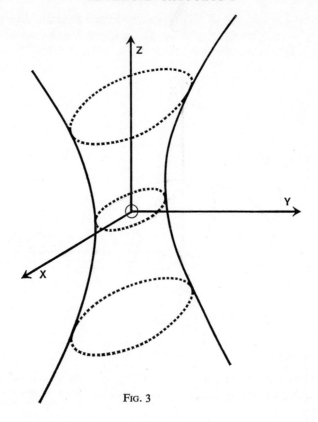

FIG. 3

(*c*) *Paraboloid*

The equations

$$z = \frac{x^2}{a^2} + \frac{y^2}{b^2} \quad \text{and} \quad z = \frac{x^2}{a^2} - \frac{y^2}{b^2}$$

represent paraboloids, the first being an elliptic paraboloid and the second a hyperbolic paraboloid. Sections parallel to the plane $z = 0$ are respectively ellipses and hyperbolas, while sections parallel to the planes $x = 0$, $y = 0$ are parabolas.

(*d*) *Cone*

The quadric may be a cone; for example, when $F(x, y, z)$ is homo-

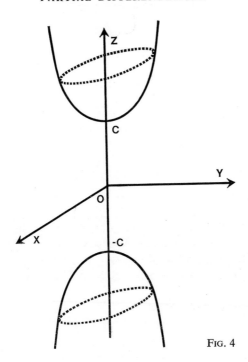

FIG. 4

geneous of degree two in x, y, z, the surface $F(x, y, z) = 0$ is a quadric cone with vertex at the origin.

(e) Cylinder

The quadric may be a cylinder; for example, when $f(x, y)$ is a polynomial of degree two in x, y, the surface $f(x, y) = 0$ is a cylinder with generators parallel to the z-axis, and sections parallel to the plane $z = 0$ are conics.

(f) Pair of planes

If $F(x, y, z)$ is a polynomial of degree two which can be factorised into two real linear factors, the quadric with equation $F(x, y, z) = 0$ is a pair of planes. For example

$$x^2 + xy + xz - 2x = 0$$

is the pair of planes $x = 0$ and $x + y + z - 2 = 0$.

WORKED EXAMPLES

1.1. *Find the nature of the surfaces represented by the equations*

   (i) $2x^2 - 2y^2 - z^2 - 4x - 2z = 0$;
   (ii) $z^3 = x^2 + y^2$;
   (iii) $y^2 + yz + z^2 = x$;
   (iv) $x^2 + y^2 - z^2 - 2z = 1$.

   (i) The equation can be written

$$(x-1)^2 - y^2 - \tfrac{1}{2}(z+1)^2 = \tfrac{1}{2},$$

showing that it is a hyperboloid of two sheets with centre at the point $(1, 0, -1)$.

   (ii) This is a surface of revolution obtained by revolving the semi-cubical parabola $z^3 = x^2$ about the z-axis.

   (iii) If we introduce by rotation new η, ζ-axes where $y = (\eta - \zeta)/\sqrt{2}$, $z = (\eta + \zeta)/\sqrt{2}$, the surface becomes $3\eta^2 + 2\zeta^2 = 2x$, showing that it is an elliptic paraboloid.

   (iv) The equation can be written

$$x^2 + y^2 - (z+1)^2 = 0,$$

showing that it represents a cone with vertex $(0, 0, -1)$.

1.2. *Find the nature of the surfaces represented by the equations*

   (i) $f(\theta, \phi) = 0$;
   (ii) $f(r, \theta) = 0$;
   (iii) $f(r, \phi) = 0$;
where r, θ, ϕ are spherical polar coordinates.

   (i) If the point $A(r, \theta, \phi)$ lies on the surface, then so do all points on the line joining A to the origin, since $f(\theta, \phi)$ is independent of r. Hence the surface is a cone with vertex the origin.

   (ii) Since the equation is independent of ϕ, if A is on the surface, then every point on the circle through A in a plane parallel to the plane $z = 0$, such that its centre lies on the z-axis also lies on the surface. Hence the surface is a surface of revolution about the z-axis.

   (iii) For a given value of ϕ, r has the same value for all values of θ. Hence if A lies on the surface, then so does every point on the circle with O as the origin, OA as radius and lying in the plane ZOA.

Thus the surface is generated by circles with centre the origin in planes passing through the z-axis, whose radii vary as the plane of the circle rotates about the z-axis.

1.3. *Find* (i) *the Cartesian coordinates of the point with spherical polar coordinates* $(2, \frac{1}{3}\pi, -\frac{1}{4}\pi)$;

   (ii) *the spherical coordinates of the point with Cartesian coordinates* $(1, -2, 3)$.

 (i) We have

$$x = r \sin \theta \cos \phi = 2 \sin \tfrac{1}{3}\pi \cos (-\tfrac{1}{4}\pi) = \sqrt{\tfrac{3}{2}},$$
$$y = r \sin \theta \sin \phi = 2 \sin \tfrac{1}{3}\pi \sin (-\tfrac{1}{4}\pi) = -\sqrt{\tfrac{3}{2}},$$
$$z = r \cos \theta = 2 \cos \tfrac{1}{3}\pi = 1.$$

The Cartesian coordinates of the point are $(\sqrt{\tfrac{3}{2}}, -\sqrt{\tfrac{3}{2}}, 1)$.

 (ii) Since

$$r \sin \theta \cos \phi = 1, \ r \sin \theta \sin \phi = -2,$$

we have $\tan \phi = -2$; that is, $\phi = \tan^{-1}(-2) = -\tan^{-1} 2$. Also, $r^2 \sin^2 \theta (\cos^2 \phi + \sin^2 \phi) = 1 + 4 = 5$, and so $r \sin \theta = \pm\sqrt{5}$. We take the positive root because $\cos \phi$ is positive and $\sin \phi$ is negative, making $r \sin \theta$ positive. Thus

$$r \sin \theta = \sqrt{5}, \ r \cos \theta = 3$$

giving $\tan \theta = \sqrt{5/3}$, and $r^2(\sin^2 \theta + \cos^2 \theta) = 5 + 9 = 14$, i.e. $r^2 = 14$. Hence the spherical polar coordinates are

$$\{\sqrt{14}, \tan^{-1}(\sqrt{5/3}), -\tan^{-1} 2\}.$$

ADDITIONAL EXAMPLES

1.4. Make a sketch of the curve, in the octant where x, y, z are positive, which is the intersection of the surfaces

$$x^2 + y^2 + z^2 = 1, \ x^2 + y^2 = y.$$

1.5. Find the equation of the cylinder with generators parallel to the z-axis which passes through the curve of intersection of the surface $x^2 + 2y^2 + z^2 = 12$, and the plane $x + y - z = 1$.

1.6. Find the equation of the cone formed by rotating the line $x = 0$, $y = 2z$ about the z-axis.

1.7. Prove that the locus of a point, the sum of whose distances from two fixed points is constant, is an ellipsoid of revolution.

SOLUTIONS

1.4.

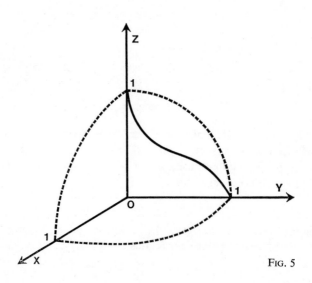

FIG. 5

1.5. $2x^2 + 3y^2 + 2xy - 2x - 2y - 11 = 0$.

1.6. $4z^2 - x^2 - y^2 = 0$.

1.7. Hint: Choose the fixed points as points on the z-axis equidistant from the origin.

2. Partial Derivatives

If in the function $z = f(x, y)$ of the two independent variables x and y, we hold y constant and differentiate with respect to x, we obtain the *partial derivative* of z at (x, y) with respect to x. This derivative is denoted by $\partial z/\partial x$ or f_x. Similarly, if we differentiate with respect to y, regarding x as a constant, we obtain the partial derivative with respect to y, written $\partial z/\partial y$ or f_y. The rules of differentiation used for finding the derivative of a function of a single variable can be used for finding partial derivatives. See **2.1**.

The partial derivatives of $f(x, y)$ with respect to x and y at the point (x_0, y_0) are defined to be respectively

$$\lim_{h \to 0} \frac{f(x_0+h, y_0)-f(x_0, y_0)}{h} \qquad (2.1)$$

and

$$\lim_{k \to 0} \frac{f(x_0, y_0+k)-f(x_0, y_0)}{k}.$$

As in the case of a function of a single variable $y = F(x)$, where the second derivative d^2y/dx^2 is defined to be $d(dy/dx)/dx$, so we define partial derivatives of the second order, thus

$$\frac{\partial^2 f}{\partial x^2} = \frac{\partial}{\partial x}\left(\frac{\partial f}{\partial x}\right), \qquad \frac{\partial^2 f}{\partial y^2} = \frac{\partial}{\partial y}\left(\frac{\partial f}{\partial y}\right), \qquad (2.2)$$

$$\frac{\partial^2 f}{\partial x \delta y} = \frac{\partial}{\partial x}\left(\frac{\partial f}{\partial y}\right), \qquad \frac{\partial^2 f}{\partial y \delta x} = \frac{\partial}{\partial y}\left(\frac{\partial f}{\partial x}\right). \qquad (2.3)$$

These derivatives are often written as $f_{xx}, f_{yy}, f_{xy}, f_{yx}$, respectively; f_{xy} and f_{yx} are called *mixed* second order partial derivatives. It can be shown that

$$\frac{\partial^2 f}{\partial x \partial y} = \frac{\partial^2 f}{\partial y \partial x}, \qquad (2.4)$$

provided that certain existence and continuity conditions are satisfied. This commutative property of partial differentiation is assumed to hold in all the problems in this chapter. Derivatives of still higher order than the second are formed in a similar way; for example,

$$f_{xxx} = \frac{\partial}{\partial x}(f_{xx}),$$

$$f_{xyx} = \frac{\partial}{\partial x}(f_{yx});$$

and in view of the commutative property the order in which the differentiations are performed does not alter the value of the derivative obtained so that $f_{xxy} = f_{xyx} = f_{yxx}$.

The concept of partial differentiation can be extended to functions of more than two variables so that if $u = f(x_1, x_2, \ldots, x_n)$, we can form $\partial u/\partial x_i \equiv f_{x_i}$, $i = 1, 2, \ldots, n$, $\partial^2 u/\partial x_i \, \partial x_j \equiv f_{x_i x_j}$, $i, j = 1, 2, \ldots, n$, and so on.

Worked Examples

2.1. *Form the first and second order partial derivatives with respect to x and y of $f(x, y) = \tan^{-1}(y/x)$ and verify that $\partial^2 f/\partial x \partial y = \partial^2 f/\partial y \partial x$.*

Here
$$f(x, y) = \tan^{-1} \frac{y}{x}. \tag{1}$$

Regarding y as a constant and differentiating (1) with respect to x, we get

$$\frac{\partial f}{\partial x} = \frac{1}{1 + \dfrac{y^2}{x^2}} \cdot \left(\frac{-y}{x^2} \right) = \frac{-y}{x^2 + y^2}, \tag{2}$$

and $\quad \dfrac{\partial^2 f}{\partial x^2} = \dfrac{\partial}{\partial x}\left(\dfrac{-y}{x^2 + y^2} \right) = -y \cdot \dfrac{-2x}{(x^2 + y^2)^2} = \dfrac{2xy}{(x^2 + y^2)^2}.$

Regarding x as a constant and differentiating (1) with respect to y, we get

$$\frac{\partial f}{\partial y} = \frac{1}{1 + \dfrac{y^2}{x^2}} \cdot \left(\frac{1}{x} \right) = \frac{x}{x^2 + y^2}, \tag{3}$$

and $\quad \dfrac{\partial^2 f}{\partial y^2} = \dfrac{\partial}{\partial y}\left(\dfrac{x}{x^2 + y^2} \right) = x \cdot \dfrac{-2y}{(x^2 + y^2)^2} = \dfrac{-2xy}{(x^2 + y^2)^2}.$

On differentiating (2) with respect to y, we get

$$\frac{\partial^2 f}{\partial y \partial x} = \frac{-1}{x^2 + y^2} + \frac{2y^2}{(x^2 + y^2)^2} = \frac{y^2 - x^2}{(x^2 + y^2)^2},$$

and on differentiating (3) with respect to x we get

$$\frac{\partial^2 f}{\partial x \partial y} = \frac{1}{x^2+y^2} - \frac{2x^2}{(x^2+y^2)^2} = \frac{y^2-x^2}{(x^2+y^2)^2}.$$

Hence we have verified that

$$\frac{\partial^2 f}{\partial y \partial x} = \frac{\partial^2 f}{\partial x \partial y}.$$

2.2. *Show that* $\log(x^2+y^2)$ *is a solution of Laplace's equation* $\nabla^2 u = 0$. $(\nabla^2 \equiv \partial^2/\partial x^2 + \partial^2/\partial y^2$ *is the two-dimensional Laplacian operator, pronounced 'del-squared'.)*

If $u = \log(x^2+y^2)$, then $\partial u/\partial x = 2x/(x^2+y^2)$,

and $\quad \dfrac{\partial^2 u}{\partial x^2} = \dfrac{2}{x^2+y^2} - \dfrac{4x^2}{(x^2+y^2)^2} = \dfrac{2(y^2-x^2)}{(x^2+y^2)^2}.$

By symmetry it follows that

$$\frac{\partial^2 u}{\partial y^2} = \frac{2(x^2-y^2)}{(x^2+y^2)^2};$$

hence $\qquad\qquad\qquad \partial^2 u/\partial x^2 + \partial^2 u/\partial y^2 = 0,$

i.e. $\qquad\qquad\qquad\qquad \nabla^2 u = 0.$

2.3. *Verify that* $u = y^{-1/2} e^{x^2/y}$ *is a solution of the equation* $\partial^2 u/\partial x^2 + 4\partial u/\partial y = 0$ *and deduce that* $u = xy^{-3/2} e^{x^2/y}$ *is also a solution.*

If $u = y^{-1/2} e^{x^2/y}$, then

$$\frac{\partial u}{\partial x} = 2xy^{-3/2} e^{x^2/y}, \frac{\partial^2 u}{\partial x^2} = (2y^{-3/2} + 4x^2 y^{-5/2}) e^{x^2/y},$$

and

$$\frac{\partial u}{\partial y} = (-\tfrac{1}{2}y^{-3/2} - x^2 y^{-5/2}) e^{x^2/y}.$$

Hence

$$\frac{\partial^2 u}{\partial x^2} + \frac{4\partial u}{\partial y} = (2y^{-3/2} + 4x^2 y^{-5/2} - 2y^{-3/2} - 4x^2 y^{-5/2}) e^{x^2/y} = 0.$$

Let $v = \dfrac{x}{y}\, u$, where $u = y^{-1/2}\, e^{x^2/y}$; then

$$\frac{\partial v}{\partial x} = \frac{u}{y} + \frac{x}{y}\frac{\partial u}{\partial x},$$

$$\frac{\partial^2 v}{\partial x^2} = \frac{1}{y}\frac{\partial u}{\partial x} + \left(\frac{1}{y}\frac{\partial u}{\partial x} + \frac{x}{y}\frac{\partial^2 u}{\partial x^2}\right) = \frac{2}{y}\frac{\partial u}{\partial x} + \frac{x}{y}\frac{\partial^2 u}{\partial x^2},$$

and

$$\frac{\partial v}{\partial y} = -\frac{x}{y^2}\, u + \frac{x}{y}\frac{\partial u}{\partial y}.$$

Hence

$$\frac{\partial^2 v}{\partial x^2} + 4\frac{\partial v}{\partial y} = \frac{2}{y}\frac{\partial u}{\partial x} + \frac{x}{y}\frac{\partial^2 u}{\partial x^2} - \frac{4x}{y^2}\, u + \frac{4x}{y}\frac{\partial u}{\partial y},$$

$$= \frac{2}{y}\frac{\partial u}{\partial x} - \frac{4x}{y^2}\, u + \frac{x}{y}\left(\frac{\partial^2 u}{\partial x^2} + 4\frac{\partial u}{\partial y}\right),$$

$$= \frac{2}{y}\frac{\partial u}{\partial x} - \frac{4x}{y^2}\, u, \quad \text{since} \quad \frac{\partial^2 u}{\partial x^2} + 4\frac{\partial u}{\partial y} = 0,$$

$$= \frac{4x}{y^{5/2}}\, e^{x^2/y} - \frac{4x}{y^{5/2}}\, e^{x^2/y} = 0.$$

A shorter method of solving the second part of this problem is to note that if $\partial^2 u/\partial x^2 + 4\partial u/\partial y = 0$, then

$$\frac{\partial^2}{\partial x^2}\left(\frac{\partial u}{\partial x}\right) + 4\frac{\partial}{\partial y}\left(\frac{\partial u}{\partial x}\right) = 0,$$

so we see at once that $2xy^{-3/2}\, e^{x^2/y}$ is also a solution.

2.4. *Prove that if* $V = f(t)$, *where* $t^2 = x^2 + y^2 + z^2$, *is a solution of* $\nabla^2 V = 0$, *where* $\nabla^2 \equiv \partial^2/\partial x^2 + \partial^2/\partial y^2 + \partial^2/\partial z^2$ *is the three-dimensional Laplacian operator, then* $f(t) = At^{-1} + B$, *where* A *and* B *are arbitrary constants.*

Since V is a function of t, where t is a function of x, we have

$$\frac{\partial V}{\partial x} = \frac{df}{dt}\frac{\partial t}{\partial x} = f'\frac{\partial t}{\partial x}.$$

But, since $t^2 = x^2+y^2+z^2$, we have $2t\partial t/\partial x = 2x$ and so $\partial V/\partial x = (x/t)f'$. On differentiating this relation partially with respect to x, we get

$$\frac{\partial^2 V}{\partial x^2} = \left(\frac{1}{t}-\frac{x}{t^2}\frac{\partial t}{\partial x}\right)f'+\frac{x}{t}f''\frac{\partial t}{\partial x}, \text{ where } f'' = \frac{d^2f}{dt^2},$$

$$= \left(\frac{1}{t}-\frac{x^2}{t^3}\right)f'+\frac{x^2}{t^2}f''.$$

From the symmetry of t in x, y, z, it follows that

$$\frac{\partial^2 V}{\partial y^2} = \left(\frac{1}{t}-\frac{y^2}{t^3}\right)f'+\frac{y^2}{t^2}f''$$

and

$$\frac{\partial^2 V}{\partial z^2} = \left(\frac{1}{t}-\frac{z^2}{t^3}\right)f'+\frac{z^2}{t^2}f''.$$

Hence

$$\nabla^2 V = \left(\frac{3}{t}-\frac{x^2+y^2+z^2}{t^3}\right)f'+\frac{x^2+y^2+z^2}{t^2}f'',$$

$$= \frac{2}{t}f'+f'', \text{ since } t^2 = x^2+y^2+z^2.$$

Now $d^2f/dt^2+(2/t)\,df/dt = 0$ is an ordinary differential equation which is solved by first writing $df/dt = \phi$, to get $\phi'+2t^{-1}\phi = 0$, so that $\phi = df/dt = (\text{constant})/t^2$, giving $f = At^{-1}+B$, where A and B are constants of integration.

2.5. *If $x^2+y^2+z^2 = a^2$ defines z as a function of x and y, show that*

$$\frac{\partial^2 z}{\partial x^2}\cdot\frac{\partial^2 z}{\partial y^2}-\left(\frac{\partial^2 z}{\partial x\partial y}\right)^2 = \frac{a^2}{z^4}.$$

Here z is an implicit function of x and y. We have $2x+2z\partial z/\partial x = 0$, so that $\partial z/\partial x = -x/z$ and similarly $\partial z/\partial y = -y/z$. Also

$$\frac{\partial^2 z}{\partial x^2} = -\frac{1}{z}+\frac{x}{z^2}\frac{\partial z}{\partial x} = -\frac{1}{z}-\frac{x^2}{z^3},$$

and similarly,

$$\frac{\partial^2 z}{\partial y^2} = -\frac{1}{z} - \frac{y^2}{z^3}.$$

Again

$$\frac{\partial^2 z}{\partial x \partial y} = \frac{\partial}{\partial x}\left(\frac{-y}{z}\right) = \frac{y}{z^2}\frac{\partial z}{\partial x} = -\frac{xy}{z^3}.$$

Thus

$$\frac{\partial^2 z}{\partial x^2} \cdot \frac{\partial^2 z}{\partial y^2} - \left(\frac{\partial^2 z}{\partial x \partial y}\right)^2 = \frac{1}{z^2} + \frac{x^2+y^2}{z^4} + \frac{x^2 y^2}{z^6} - \frac{x^2 y^2}{z^6},$$

$$= \frac{x^2 + y^2 + z^2}{z^4}$$

$$= \frac{a^2}{z^4}.$$

ADDITIONAL EXAMPLES

2.6. Verify that $f = \sin^{-1}(y/x)/(x^2 + y^2)$ satisfies the equation

$$x\frac{\partial f}{\partial x} + y\frac{\partial f}{\partial y} + 2f = 0.$$

2.7. If $u = (1 - 2xy + y^2)^{-1/2}$, prove that

$$\frac{\partial}{\partial x}\left((1-x^2)\frac{\partial u}{\partial x}\right) + \frac{\partial}{\partial y}\left(y^2\frac{\partial u}{\partial y}\right) = 0.$$

2.8. If $\log V = at + bx^2/t - \log Vt$, prove that

$$\frac{\partial^2 V}{\partial x^2} + 2b\frac{\partial V}{\partial t} = abV.$$

2.9. If $V = \phi(u)$, where $u = xyz$, prove that

$$\frac{\partial^3 V}{\partial x \partial y \partial z} = u^2 \phi''' + 3u\phi'' + \phi'.$$

2.10. If $a^2x^2 + b^2y^2 = c^2z^2$, show that

$$\frac{1}{a^2}\frac{\partial^2 z}{\partial x^2} + \frac{1}{b^2}\frac{\partial^2 z}{\partial y^2} = \frac{1}{c^2 z}.$$

2.11. Show that $\phi = Ae^{-\frac{1}{2}kt}\cos mx \cos nt$ is a solution of the equation

$$c^2\frac{\partial^2 \phi}{\partial x^2} = \frac{\partial^2 \phi}{\partial t^2} + k\frac{\partial \phi}{\partial t},$$

provided that the constants k, m, n, c are related by the equation $n^2 + \frac{1}{4}k^2 = c^2m^2$.

2.12. If $z = f(x+y)$ satisfies the equation $x\partial z/\partial x + y\partial z/\partial y = 1$, show that $(x+y)f'(x+y) = 1$ and deduce the form of the function f.

2.13. Show that, if f and g are any functions, $u = \{f(x-ct) + g(x+ct)\}/x$ satisfies the equation

$$\frac{1}{c^2}\frac{\partial^2 u}{\partial t^2} = \frac{\partial^2 u}{\partial x^2} + \frac{2}{x}\frac{\partial u}{\partial x}.$$

Deduce that $u = \{\sin(x/a)\cos(ct/a)\}/x$ is a solution

(*i*) by direct substitution in the equation;
(*ii*) by expressing it in the form of the above general equation.

2.14. Obtain the second order partial differential equation satisfied by V, which arises through the elimination of the functions $f(x)$ and $\phi(y)$ from the relation $V = x^2\phi(y) + y^2f(x)$. Find a solution of this differential equation which is such that $V = x^6, \partial V/\partial x = 0$ when $y = x$.

2.15. Show that, if f and g are any functions, $z = f(xe^y) + g(xe^{-y})$ satisfies the equation

$$x^2\frac{\partial^2 z}{\partial x^2} + x\frac{\partial z}{\partial x} - \frac{\partial^2 z}{\partial y^2} = 0.$$

SOLUTIONS

2.12. $f(t) = \log t + A$, where A is an arbitrary constant.

2.14. $xy \dfrac{\partial^2 V}{\partial x \partial y} - 2\left(y\dfrac{\partial V}{\partial y} + x\dfrac{\partial V}{\partial x}\right) + 4V = 0$, $V = 2x^2y^4 - x^4y^2$.

3. Change of Variable: Chain Rule

Let $z = f(x, y)$ be a function of two variables x, y, where $x = x(t)$, $y = y(t)$ are functions of the single independent variable t. Then z is a function of t, say $z = F(t)$ and it can be shown that subject to certain existence and continuity conditions

$$\frac{dz}{dt} = \frac{\partial z}{\partial x}\frac{dx}{dt} + \frac{\partial z}{\partial y}\frac{dy}{dt}. \tag{3.1}$$

This result can be extended to the case where
$z = f(x_1, \ldots, x_n)$, $x_1 = x_1(t), \ldots, x_n = x_n(t)$, when we have

$$\frac{dz}{dt} = \sum_{r=1}^{n} \frac{\partial z}{\partial x_r}\frac{dx_r}{dt}. \tag{3.2}$$

Again, if $z = f(x, y)$, where $x = x(u, v)$, $y = y(u, v)$, z may be considered as a function of u, v and we have

$$\frac{\partial z}{\partial u} = \frac{\partial z}{\partial x}\frac{\partial x}{\partial u} + \frac{\partial z}{\partial y}\frac{\partial y}{\partial u}, \qquad \frac{\partial z}{\partial v} = \frac{\partial z}{\partial x}\frac{\partial x}{\partial v} + \frac{\partial z}{\partial y}\frac{\partial y}{\partial v}. \tag{3.3}$$

In this case, u, v may be taken as the intermediate variables with x, y as the independent variables and we have

$$\frac{\partial z}{\partial x} = \frac{\partial z}{\partial u}\frac{\partial u}{\partial x} + \frac{\partial z}{\partial v}\frac{\partial v}{\partial x}, \qquad \frac{\partial z}{\partial y} = \frac{\partial z}{\partial u}\frac{\partial u}{\partial y} + \frac{\partial z}{\partial v}\frac{\partial v}{\partial y} \tag{3.4}$$

There are corresponding results for $z = f(x_1, \ldots, x_n)$, where $x_i = x_i(t_1, \ldots, t_m)$, $i = 1, \ldots, n$. We have in fact

$$\frac{\partial z}{\partial t_j} = \sum_{i=1}^{n} \frac{\partial z}{\partial x_i}\frac{\partial x_i}{\partial t_j}, \qquad j = 1, \ldots, m. \tag{3.5}$$

Higher order derivatives are found by repeated application of these rules. The method is illustrated in some of the following examples.

WORKED EXAMPLES

3.1. *The volume of a circular cylinder is increasing at the rate of* 2 *cm*^{3}/*s and its height is increasing at* 0·04 *cm*/*s. Find the rate at which the radius is changing when the radius is* 6 *cm and the height is* 4 *cm.*

Let V cm^3 be the volume, r cm the radius and h cm the height at time t s. Then $V = \pi r^2 h$, and application of (*3.1*) gives

$$\frac{dV}{dt} = 2\pi r h \frac{dr}{dt} + \pi r^2 \frac{dh}{dt}. \tag{1}$$

Substitution of $r = 6$, $h = 4$, $dV/dt = 2$, $dh/dt = 0·04$ gives $dr/dt = -0·017$, so that the radius is decreasing at the rate of 0·017 cm/s.

3.2. *When* $x = u+v$, $y = uv$ *are substituted for* x, y *in* $f(x, y)$, $f(x, y)$ *becomes* $F(u, v)$. *Show that*

$$\frac{\partial f}{\partial x} = \frac{1}{u-v}\left(u\frac{\partial F}{\partial u} - v\frac{\partial F}{\partial v}\right), \qquad \frac{\partial f}{\partial y} = \frac{1}{u-v}\left(\frac{\partial F}{\partial v} - \frac{\partial F}{\partial u}\right).$$

Since $x = u+v$, $y = uv$, we have $\partial x/\partial u = 1$, $\partial x/\partial v = 1$, $\partial y/\partial u = v$, $\partial y/\partial v = u$ and substituting these values in (*3.3*) gives

$$\frac{\partial F}{\partial u} = \frac{\partial f}{\partial x} + v\frac{\partial f}{\partial y}, \qquad \frac{\partial F}{\partial v} = \frac{\partial f}{\partial x} + u\frac{\partial f}{\partial y}.$$

Solving these equations for $\partial f/\partial x$ and $\partial f/\partial y$ gives the required relations.

3.3. *If* $z = f(x, y)$, *where* $x = x(t)$, $y = y(t)$, *obtain an expression for* d^2z/dt^2.

From (*3.1*) $$\frac{dz}{dt} = \frac{\partial z}{\partial x}\frac{dx}{dt} + \frac{\partial z}{\partial y}\frac{dy}{dt},$$

so that, operationally,

$$\frac{d}{dt} = \frac{dx}{dt}\frac{\partial}{\partial x} + \frac{dy}{dt}\frac{\partial}{\partial y}. \tag{1}$$

Now

$$\frac{d^2z}{dt^2} = \frac{d}{dt}\left(\frac{dz}{dt}\right)$$

$$= \frac{dx}{dt}\frac{d}{dt}\left(\frac{\partial z}{\partial x}\right) + \frac{\partial z}{\partial x}\frac{d^2x}{dt^2} + \frac{dy}{dt}\frac{d}{dt}\left(\frac{\partial z}{\partial y}\right) + \frac{\partial z}{\partial y}\frac{d^2y}{dt^2}$$

$$= \left\{\left(\frac{dx}{dt}\frac{\partial}{\partial x} + \frac{dy}{dt}\frac{\partial}{\partial y}\right)\frac{\partial z}{\partial x}\right\}\frac{dx}{dt} + \frac{\partial z}{\partial x}\frac{d^2x}{dt^2}$$

$$+ \left\{\left(\frac{dx}{dt}\frac{\partial}{\partial x} + \frac{dy}{dt}\frac{\partial}{\partial y}\right)\frac{\partial z}{\partial y}\right\}\frac{dy}{dt} + \frac{\partial z}{\partial y}\frac{d^2y}{dt^2}$$

$$= \frac{\partial^2 z}{\partial x^2}\left(\frac{dx}{dt}\right)^2 + 2\frac{\partial^2 z}{\partial x\partial y}\left(\frac{dx}{dt}\right)\left(\frac{dy}{dt}\right) + \frac{\partial^2 z}{\partial y^2}\left(\frac{dy}{dt}\right)^2 + \frac{\partial z}{\partial x}\frac{d^2x}{dt^2} + \frac{\partial z}{\partial y}\frac{d^2y}{dt^2}.$$

3.4. *If f is a function of x, y and* $u = x^m y^n$, $v = x^n y^{-m}$, *prove that*

(i) $x^2\left(\dfrac{\partial f}{\partial x}\right)^2 + y^2\left(\dfrac{\partial f}{\partial y}\right)^2 = (m^2+n^2)\left\{u^2\left(\dfrac{\partial f}{\partial u}\right)^2 + v^2\left(\dfrac{\partial f}{\partial v}\right)^2\right\};$

(ii) $x^2\dfrac{\partial^2 f}{\partial x^2} + y^2\dfrac{\partial^2 f}{\partial y^2} + x\dfrac{\partial f}{\partial x} + y\dfrac{\partial f}{\partial y}$

$$= (m^2+n^2)\left(u^2\frac{\partial^2 f}{\partial u^2} + v^2\frac{\partial^2 f}{\partial v^2} + u\frac{\partial f}{\partial u} + v\frac{\partial f}{\partial v}\right).$$

Using (*3.4*) we have

$$\frac{\partial f}{\partial x} = mx^{m-1}y^n\frac{\partial f}{\partial u} + nx^{n-1}y^{-m}\frac{\partial f}{\partial v},$$

$$\frac{\partial f}{\partial y} = nx^m y^{n-1}\frac{\partial f}{\partial u} - mx^n y^{-m-1}\frac{\partial f}{\partial v},$$

so that

(i) $x^2\left(\dfrac{\partial f}{\partial x}\right)^2 + y^2\left(\dfrac{\partial f}{\partial y}\right)^2 = \left(mu\dfrac{\partial f}{\partial u} + nv\dfrac{\partial f}{\partial v}\right)^2 + \left(nu\dfrac{\partial f}{\partial u} - mv\dfrac{\partial f}{\partial v}\right)^2$

$$= (m^2+n^2)\left\{u^2\left(\frac{\partial f}{\partial u}\right)^2 + v^2\left(\frac{\partial f}{\partial v}\right)^2\right\}.$$

Now $\dfrac{\partial}{\partial x} = mx^{m-1}y^n\dfrac{\partial}{\partial u} + nx^{n-1}y^{-m}\dfrac{\partial}{\partial v}$, so that

$$\frac{\partial^2 f}{\partial x^2} = \frac{\partial}{\partial x}\left(\frac{\partial f}{\partial x}\right) = m(m-1)x^{m-2}y^n\frac{\partial f}{\partial u}$$

$$+ mx^{m-1}y^n\left(mx^{m-1}y^n\frac{\partial^2 f}{\partial u^2} + nx^{n-1}y^{-m}\frac{\partial^2 f}{\partial u\partial v}\right)$$

$$+ n(n-1)x^{n-2}y^{-m}\frac{\partial f}{\partial v}$$

$$+ nx^{n-1}y^{-m}\left(mx^{m-1}y^n\frac{\partial^2 f}{\partial u\partial v} + nx^{n-1}y^{-m}\frac{\partial^2 f}{\partial v^2}\right)$$

$$= m^2 x^{2m-2}y^{2n}\frac{\partial^2 f}{\partial u^2} + 2mnx^{m+n-2}y^{n-m}\frac{\partial^2 f}{\partial u\partial v} + n^2 x^{2n-2}y^{-2m}\frac{\partial^2 f}{\partial v^2}$$

$$+ m(m-1)x^{m-2}y^n\frac{\partial f}{\partial u} + n(n-1)x^{n-2}y^{-m}\frac{\partial f}{\partial v}.$$

By similar reasoning we find that

$$\frac{\partial^2 f}{\partial y^2} = n^2 x^{2m}y^{2n-2}\frac{\partial^2 f}{\partial u^2} - 2mnx^{m+n}y^{n-m-2}\frac{\partial^2 f}{\partial u\partial v} + m^2 x^{2n}y^{-2m-2}\frac{\partial^2 f}{\partial v^2}$$

$$+ n(n-1)x^m y^{n-2}\frac{\partial f}{\partial u} + m(m+1)x^n y^{-m-2}\frac{\partial f}{\partial v}.$$

Hence

(ii) $x^2 f_{xx} + y^2 f_{yy} + xf_x + yf_y = (m^2 + n^2)(u^2 f_{uu} + v^2 f_{vv} + uf_u + vf_v).$

3.5. *If u, v are functions of x, y such that $u_x = v_y$, $v_x = -u_y$, show that both u and v are harmonic functions. Show also that if f is a function of x, y (and so of u, v) then*

(i) $f_x^2 + f_y^2 = (u_x^2 + u_y^2)(f_u^2 + f_v^2);$

(ii) $f_{xx} + f_{yy} = (u_x^2 + u_y^2)(f_{uu} + f_{vv}).$

Functions satisfying Laplace's equation $\nabla^2\phi = 0$, where $\nabla^2 \equiv \partial^2/\partial x^2 + \partial^2/\partial y^2$, are called *harmonic* functions. We are given here

$$\frac{\partial u}{\partial x} = \frac{\partial v}{\partial y}, \tag{1}$$

$$\frac{\partial v}{\partial x} = -\frac{\partial u}{\partial y}; \tag{2}$$

these are the *Cauchy-Riemann equations.*

Differentiation of (*1*) with respect to x and of (*2*) with respect to y gives

$$\frac{\partial^2 u}{\partial x^2} = \frac{\partial^2 v}{\partial x \partial y} = \frac{\partial^2 v}{\partial y \partial x} = -\frac{\partial^2 u}{\partial y^2},$$

and so $u_{xx} + u_{yy} = 0$, i.e., u is a harmonic function. Similarly, differentiation of (*1*) with respect y and of (*2*) with respect to x gives $v_{xx} + v_{yy} = 0$, so that v is also a harmonic function.

Now
$$\frac{\partial f}{\partial x} = \frac{\partial f}{\partial u}\frac{\partial u}{\partial x} + \frac{\partial f}{\partial v}\frac{\partial v}{\partial x} = \frac{\partial f}{\partial u}\frac{\partial u}{\partial x} - \frac{\partial f}{\partial v}\frac{\partial u}{\partial y}, \tag{3}$$

$$\frac{\partial f}{\partial y} = \frac{\partial f}{\partial u}\frac{\partial u}{\partial y} + \frac{\partial f}{\partial v}\frac{\partial v}{\partial y} = \frac{\partial f}{\partial u}\frac{\partial u}{\partial y} + \frac{\partial f}{\partial v}\frac{\partial u}{\partial x}. \tag{4}$$

Squaring and adding (*3*) and (*4*) we get

$$f_x^2 + f_y^2 = (u_x^2 + u_y^2)(f_u^2 + f_v^2).$$

Note that $\partial/\partial x = (\partial u/\partial x)\partial/\partial u - (\partial u/\partial y)\partial/\partial v$, so that, as in the previous example,

$$\frac{\partial^2 f}{\partial x^2} = \frac{\partial u}{\partial x}\frac{\partial}{\partial x}\left(\frac{\partial f}{\partial u}\right) + \frac{\partial f}{\partial u}\frac{\partial^2 u}{\partial x^2} - \frac{\partial u}{\partial y}\frac{\partial}{\partial x}\left(\frac{\partial f}{\partial v}\right) - \frac{\partial f}{\partial v}\frac{\partial^2 u}{\partial x \partial y}$$

$$= \left(\frac{\partial^2 f}{\partial u^2}\frac{\partial u}{\partial x} - \frac{\partial^2 f}{\partial v \partial u}\frac{\partial u}{\partial y}\right)\frac{\partial u}{\partial x} + \frac{\partial f}{\partial u}\frac{\partial^2 u}{\partial x^2} - \left(\frac{\partial^2 f}{\partial v \partial u}\frac{\partial u}{\partial x} - \frac{\partial^2 f}{\partial v^2}\frac{\partial u}{\partial y}\right)\frac{\partial u}{\partial y}$$

$$- \frac{\partial f}{\partial v}\frac{\partial^2 u}{\partial x \partial y}$$

$$= \frac{\partial^2 f}{\partial u^2}\left(\frac{\partial u}{\partial x}\right)^2 - 2\frac{\partial^2 f}{\partial u \partial v}\left(\frac{\partial u}{\partial x}\right)\left(\frac{\partial u}{\partial y}\right) + \frac{\partial^2 f}{\partial v^2}\left(\frac{\partial u}{\partial y}\right)^2 + \frac{\partial f}{\partial u}\frac{\partial^2 u}{\partial x^2} - \frac{\partial f}{\partial v}\frac{\partial^2 u}{\partial x \partial y}.$$

Similarly it may be shown that

$$\frac{\partial^2 f}{\partial y^2} = \frac{\partial^2 f}{\partial u^2}\left(\frac{\partial u}{\partial y}\right)^2 + 2\frac{\partial^2 f}{\partial u \partial v}\left(\frac{\partial u}{\partial x}\right)\left(\frac{\partial u}{\partial y}\right) + \frac{\partial^2 f}{\partial v^2}\left(\frac{\partial u}{\partial x}\right)^2 + \frac{\partial f}{\partial u}\frac{\partial^2 u}{\partial y^2} + \frac{\partial f}{\partial v}\frac{\partial^2 u}{\partial x \partial y}.$$

Since $u_{xx} + u_{yy} = 0$, it follows that

$$f_{xx} + f_{yy} = (u_x^2 + u_y^2)(f_{uu} + f_{vv}).$$

3.6. *Show that if* $V = f(x+ct) + g(x-ct)$, *then* $\partial^2 V/\partial t^2 = c^2 \partial^2 V/\partial x^2$ *and conversely any solution of the partial differential equation has the form of* V.

The equation
$$\frac{\partial^2 V}{\partial t^2} = c^2 \frac{\partial^2 V}{\partial x^2} \tag{1}$$

is the wave equation. If

$$V = f(x+ct) + g(x-ct) \tag{2}$$

we have
$$\frac{\partial V}{\partial x} = f'(x+ct) + g'(x-ct),$$

$$\frac{\partial^2 V}{\partial x^2} = f''(x+ct) + g''(x-ct),$$

$$\frac{\partial V}{\partial t} = cf'(x+ct) - cg'(x-ct),$$

$$\frac{\partial^2 V}{\partial t^2} = c^2 f''(x+ct) + c^2 g''(x-ct),$$

so that $\partial^2 V/\partial t^2 = c^2 \partial^2 V/\partial x^2$.

Conversely, suppose V satisfies (*1*). Let $\xi = x+ct$, $\eta = x-ct$, so that $x = (\xi+\eta)/2$, $t = (\xi-\eta)/2c$. Hence we can consider V as a function of ξ, η. We have

$$\frac{\partial V}{\partial \eta} = \frac{1}{2}\frac{\partial V}{\partial x} - \frac{1}{2c}\frac{\partial V}{\partial t},$$

$$\frac{\partial^2 V}{\partial \xi \partial \eta} = \frac{1}{2}\left(\frac{1}{2}\frac{\partial^2 V}{\partial x^2} + \frac{1}{2c}\frac{\partial^2 V}{\partial t \partial x}\right) - \frac{1}{2c}\left(\frac{1}{2}\frac{\partial^2 V}{\partial x \partial t} + \frac{1}{2c}\frac{\partial^2 V}{\partial t^2}\right)$$

$$= \frac{1}{4}\left(\frac{\partial^2 V}{\partial x^2} - \frac{1}{c^2}\frac{\partial^2 V}{\partial t^2}\right) = 0,$$

since V satisfies (1). Since $\partial(\partial V/\partial\eta)/\partial\xi = 0$, it follows that

$$\frac{\partial V}{\partial \eta} = \text{a function of } \eta = \lambda(\eta), \text{ say,}$$

where $\lambda(\eta)$ is an arbitrary function of η. Again it follows that

$$V = \int \lambda(\eta)d\eta + \text{a function of } \xi$$

$$= g(\eta) + f(\xi),$$

where f and g are arbitrary functions. Hence

$$V = f(x+ct) + g(x-ct).$$

ADDITIONAL EXAMPLES

3.7. If x increases at the rate of 2 cm/s at the instant when $x = 3$ cm and $y = 1$ cm, find the rate at which y must be changing in order that the function $2xy - 3x^2y$ shall be neither increasing nor decreasing when x and y have these values.

3.8. If ϕ is a function of x and y, and $u = x+y$, $v = x^{-1}+y^{-1}$, show that

$$x\frac{\partial\phi}{\partial x} + y\frac{\partial\phi}{\partial y} = u\frac{\partial\phi}{\partial u} - v\frac{\partial\phi}{\partial v}.$$

3.9. u, v are functions of x, y which satisfy the Cauchy-Riemann equations $u_x = v_y$, $v_x = -u_y$. If u, v are expressed in terms of r, θ by means of the relations $x = r\cos\theta$, $y = r\sin\theta$, show that

$$\frac{\partial u}{\partial r} = \frac{1}{r}\frac{\partial v}{\partial \theta}, \qquad \frac{\partial v}{\partial r} = -\frac{1}{r}\frac{\partial u}{\partial \theta},$$

and hence show that both u and v satisfy the equation

$$\frac{\partial^2\phi}{\partial r^2} + \frac{1}{r}\frac{\partial\phi}{\partial r} + \frac{1}{r^2}\frac{\partial^2\phi}{\partial\theta^2} = 0.$$

3.10. By changing the independent variables to u, v, where $u = \frac{1}{2}\log(x^2+y^2)$, $v = \tan^{-1}(y/x)$, show that the general solution of the equation

$$x\frac{\partial z}{\partial x}+y\frac{\partial z}{\partial y} = x^2+y^2$$

is $z = \frac{1}{2}(x^2+y^2)+f(\tan^{-1}(y/x))$, where f is an arbitrary function.

3.11. If the independent variables, x, y are changed to u, v, where $xy = u$, $xv = y$, show that, operationally,

$$x\frac{\partial}{\partial x} \equiv u\frac{\partial}{\partial u}-v\frac{\partial}{\partial v}, \quad y\frac{\partial}{\partial y} \equiv u\frac{\partial}{\partial u}+v\frac{\partial}{\partial v},$$

and solve the equation

$$x^2\frac{\partial^2 z}{\partial x^2}-y^2\frac{\partial^2 z}{\partial y^2}+x\frac{\partial z}{\partial x}-y\frac{\partial z}{\partial y} = 0$$

by changing the independent variables to u, v.

3.12. Show that when the equation

$$\frac{\partial^2 F}{\partial x^2}+2xy^2\frac{\partial F}{\partial x}+2(y-y^3)\frac{\partial F}{\partial y}+x^2y^2F = 0$$

is transformed by the substitutions, $u = xy$, $vy = 1$, the same equation is obtained with u, v in place of x, y.

3.13. Show that if $x^2\partial^2 z/\partial x^2 - \partial^2 z/\partial y^2 + x\partial z/\partial x = 0$, then $z = f(xe^y)+g(xe^{-y})$.

3.14. If $u(x, y)$ becomes $f(r, \theta)$ when x is replaced by $r\cos\theta$ and y by $r\sin\theta$, show that

$$r^2\frac{\partial^2 f}{\partial r^2} = x^2\frac{\partial^2 u}{\partial x^2}+2xy\frac{\partial^2 u}{\partial x\partial y}+y^2\frac{\partial^2 u}{\partial y^2},$$

and that, for any positive integer n,

$$r^n\frac{\partial^n f}{\partial r^n} = x^n\frac{\partial^n u}{\partial x^n}+ \ldots +\binom{n}{k}x^{n-k}y^k\frac{\partial^n u}{\partial x^{n-k}\partial y^k}+ \ldots +y^n\frac{\partial^n u}{\partial y^n}.$$

3.15. The function $z(x, y)$ is transformed into a function of u, v by means of the substitution $x = e^u \cos v$, $y = e^u \sin v$. Show that if z is a harmonic function in x, y, it is also a harmonic function in u, v.

3.16. Show that the only solution of Laplace's equation $\partial^2 u/\partial x^2 + \partial^2 u/\partial y^2 + \partial^2 u/\partial z^2 = 0$ which is of the form $u = f(r+z)$ where $r^2 = x^2 + y^2 + z^2$ is $u = A \log(r+z) + B$.

Solutions

3.7. y is decreasing at the rate of $32/21$ cm/s.

3.11. $z = f(xy) + g(y/x)$, where f and g are arbitrary functions.

4. Taylor's Series: Tangent Plane: Directional Derivatives

(*i*) Taylor's theorem for a function of two variables $f(x, y)$ states that under certain continuity conditions

$$f(x+h, y+k) = f(x, y) + \left(h \frac{\partial}{\partial x} + k \frac{\partial}{\partial y} \right) f(x, y)$$

$$+ \frac{1}{2!} \left(h \frac{\partial}{\partial x} + k \frac{\partial}{\partial y} \right)^2 f(x, y) + \ldots$$

$$+ \frac{1}{n!} \left(h \frac{\partial}{\partial x} + k \frac{\partial}{\partial y} \right)^n f(x, y) + R_n \qquad (4.1)$$

where $\qquad R_n = \frac{1}{(n+1)!} \left(h \frac{\partial}{\partial x} + k \frac{\partial}{\partial y} \right)^{n+1} f(x+\theta h, y+\theta k),$

$$0 < \theta < 1.$$

Here

$$\left(h \frac{\partial}{\partial x} + k \frac{\partial}{\partial y} \right)^2 f = \left(h \frac{\partial}{\partial x} + k \frac{\partial}{\partial y} \right) \left(h \frac{\partial f}{\partial x} + k \frac{\partial f}{\partial y} \right)$$

$$= h^2 \frac{\partial^2 f}{\partial x^2} + 2hk \frac{\partial^2 f}{\partial x \partial y} + k^2 \frac{\partial^2 f}{\partial y^2}, \text{ etc.}$$

Taylor's theorem can be extended to functions of any number of variables, e.g.,

$$f(x+h, y+k, z+l) = f(x, y, z) + \sum_{r=1}^{n} \left(h\frac{\partial}{\partial x} + k\frac{\partial}{\partial y} + l\frac{\partial}{\partial z} \right)^r f(x, y, z)$$
$$+ R_n, \qquad (4.2)$$

where $R_n = \dfrac{1}{(n+1)!} \left(h\dfrac{\partial}{\partial x} + k\dfrac{\partial}{\partial y} + l\dfrac{\partial}{\partial z} \right)^{n+1} f(x+\theta h, y+\theta k, z+\theta l),$
$$0 < \theta < 1.$$

(*ii*) By means of Taylor's theorem, it can be shown that the normal to the surface $f(x, y, z) = 0$ at the point (α, β, γ) on it has direction cosines proportional to $(f_\alpha, f_\beta, f_\gamma)$, where $f_\alpha, f_\beta, f_\gamma$ denote the values of f_x, f_y, f_z at (α, β, γ), and that the tangent plane at the point (α, β, γ) has equation

$$(x-\alpha)f_\alpha + (y-\beta)f_\beta + (z-\gamma)f_\gamma = 0. \qquad (4.3)$$

If the surface is given by the equation $z = z(x, y)$, the normal at (α, β, γ) has direction cosines proportional to $(p, q, -1)$ where p, q denote the values of $\partial z/\partial x$, $\partial z/\partial y$ at (α, β, γ), and the tangent plane at (α, β, γ) has equation

$$(x-\alpha)p + (y-\beta)q = z-\gamma. \qquad (4.4)$$

(*iii*) Let u be a function of x, y, where x, y are Cartesian co-ordinates of a point in a plane. The rate of change of u in a direction making an angle θ with the positive x-axis, or the derivative of u in that direction can be shown to be

$$\cos\theta \frac{\partial u}{\partial x} + \sin\theta \frac{\partial u}{\partial y}. \qquad (4.5)$$

Similarly, if $u = u(x, y, z)$, where x, y, z are Cartesian coordinates in space, the derivative of u in a direction with direction cosines (l, m, n) can be shown to be

$$l\frac{\partial u}{\partial x} + m\frac{\partial u}{\partial y} + n\frac{\partial u}{\partial z}. \qquad (4.6)$$

B

WORKED EXAMPLES

4.1. *Show that*

$$f(x+h, y+k)-f(x, y) = \left(h\,\frac{\partial}{\partial x}+k\,\frac{\partial}{\partial y}\right)f(x_1, y_1)$$

where (x_1, y_1) *is a point lying on the line joining* (x, y) *and* $(x+h, y+k)$.

In (*4.1*), take $n = 0$. The point $(x+\theta h, y+\theta k)$, $0 < \theta < 1$, lies on the line joining (x, y) and $(x+h, y+k)$ and between them. This is the *Mean Value Theorem* for a function of two variables.

4.2. *If* $f(x, y)$ *is homogeneous of degree n in x and y, show that*

(*i*) $\quad x\,\dfrac{\partial f}{\partial x}+y\,\dfrac{\partial f}{\partial y} = nf;$

(*ii*) $\quad \displaystyle\sum_{m=0}^{r}\binom{r}{m}x^{r-m}y^{m}\frac{\partial^{r}f}{\partial x^{r-m}\partial y^{m}} = n(n-1)\ldots(n-r+1)f.$

These well known results are *Euler's Theorems on homogeneous functions*.

Since f is homogeneous of degree n, we have [as in §1 (*iv*)]

$$f(\lambda x, \lambda y) = \lambda^{n}f(x, y).$$

Take $\lambda = 1+t$, so that

$$f(x+xt, y+yt) = (1+t)^{n}f(x, y).$$

We now expand the left-hand side by Taylor's theorem and the right-hand side by the binomial theorem:

$$f(x, y)+t\left(x\,\frac{\partial f}{\partial x}+y\,\frac{\partial f}{\partial y}\right)+\frac{1}{2!}\,t^{2}\left(x^{2}\,\frac{\partial^{2}f}{\partial x^{2}}+2xy\,\frac{\partial^{2}f}{\partial x\partial y}+y^{2}\,\frac{\partial^{2}f}{\partial y^{2}}\right)+\cdots$$

$$+\frac{1}{r!}\,t^{r}\left(\sum_{m=0}^{r}\binom{r}{m}x^{r-m}y^{m}\frac{\partial^{2}f}{\partial x^{r-m}\partial y^{r}}\right)+\cdots$$

$$=\left\{1+nt+\frac{n(n-1)}{2!}\,t^{2}+\cdots+\frac{n(n-1)\ldots(n-r+1)}{r!}\,t^{r}+\cdots\right\}$$

$$\times f(x, y).$$

We get the results (*i*) and (*ii*) by equating coefficients of t and t^{r}.

The converse results also hold. If $xf_x + yf_y = nf$, then (ii) can be proved by induction, and the argument given above reversed so that f is homogeneous of degree n in x, y.

The results can clearly be extended to functions of any number of variables.

4.3. Let $f(x, y, z) = 0$ be the equation of a surface, and let $F(x, y, z, t) \equiv t^n f(x/t, y/t, z/t)$. Show that F is homogeneous of degree n in x, y, z, t and that the equation of the tangent plane at the point (α, β, γ) on $f(x, y, z) = 0$ is

$$x F_\alpha + y F_\beta + z F_\gamma + F_t = 0,$$

where $F_\alpha, F_\beta, F_\gamma, F_t$ denote the values of F_x, F_y, F_z, F_t at $(\alpha, \beta, \gamma, 1)$.

Since $F(\lambda x, \lambda y, \lambda z, \lambda t) = (\lambda t)^n f\left(\dfrac{\lambda x}{\lambda t}, \dfrac{\lambda y}{\lambda t}, \dfrac{\lambda z}{\lambda t}\right) = \lambda^n F(x, y, z, t)$, F is

clearly homogeneous of degree n in x, y, z, t. Now
$F(x, y, z, 1) = f(x, y, z)$, and

$$\left[\frac{\partial F(x, y, z, t)}{\partial x}\right]_{t=1} = \frac{\partial F(x, y, z, 1)}{\partial x} = \frac{\partial f(x, y, z)}{\partial x},$$

since t is constant during this differentiation, and therefore giving t the particular value 1 after differentiation yields the same result as doing so before differentiation. We can treat f_y and f_z similarly. Hence from (4.3), the equation of the tangent plane at (α, β, γ) becomes

$$(x - \alpha) F_\alpha + (y - \beta) F_\beta + (z - \gamma) F_\gamma = 0.$$

Further, since F is homogeneous of degree n,

$$\alpha F_\alpha + \beta F_\beta + \gamma F_\gamma + F_t = nF(\alpha, \beta, \gamma, 1) = nf(\alpha, \beta, \gamma) = 0,$$

and it follows that the equation of the tangent plane at (α, β, γ) takes the form

$$x F_\alpha + y F_\beta + z F_\gamma + F_t = 0.$$

4.4. Find the equation of the tangent plane at the point $(1, -1, 3)$ on the quadric surface

$$x^2 + y^2 - yz + 2x - 3z + 2 = 0.$$

We use the method and notation of the last example:

$$F = x^2 + y^2 - yz + 2xt - 3zt + 2t^2,$$

$$F_x = 2x + 2t, \ F_y = 2y - z, \ F_z = -y - 3t, \ F_t = 2x - 3z + 4t,$$

so that when $x = 1$, $y = -1$, $z = 3$, $t = 1$, $F_x = 4$, $F_y = -5$, $F_z = -2$, $F_t = -3$. Hence the required tangent plane is

$$4x - 5y - 2z - 3 = 0.$$

4.5. *Find the derivative of the function* $x^2/a^2 + y^2/b^2$ *at the point* (x, y) *on the ellipse* $x^2/a^2 + y^2/b^2 = 1$, *in the direction of the outward normal to the ellipse.*

For the ellipse $dy/dx = -b^2x/a^2y$, so that the gradient of the normal is a^2y/b^2x. Hence, if the normal makes angle θ with the positive x-axis

$$\cos\theta = \frac{b^2x}{\sqrt{(b^4x^2 + a^4y^2)}}, \quad \sin\theta = \frac{a^2y}{\sqrt{(b^4x^2 + a^4y^2)}}.$$

Thus the derivative of $x^2/a^2 + y^2/b^2$ in the direction of the outward normal is

$$\frac{2\dfrac{b^2x^2}{a^2} + 2\dfrac{a^2y^2}{b^2}}{\sqrt{(b^4x^2 + a^4y^2)}} = \frac{2}{a^2b^2}\sqrt{(b^4x^2 + a^4y^2)}.$$

4.6. *If* $u = u(r, \theta)$, *when* r, θ *are polar coordinates, show that the derivative of* u *at the point* (r, θ) *in the direction making angle* ψ *with the direction of the radius vector from the origin to the point* (r, θ) *is*

$$\cos\psi\,\frac{\partial u}{\partial r} + \frac{\sin\psi}{r}\,\frac{\partial u}{\partial\theta}.$$

Since $x = r\cos\theta$, $y = r\sin\theta$, we have $u_r = u_x\cos\theta + u_y\sin\theta$, $u_\theta = u_x(-r\sin\theta) + u_y(r\cos\theta)$, whence we obtain

$$u_x = \cos\theta\,u_r - \frac{\sin\theta}{r}\,u_\theta, \ u_y = \sin\theta\,u_r + \frac{\cos\theta}{r}\,u_\theta$$

Now the required direction makes angle $\theta + \psi$ with the x-axis, and so the derivative in this direction is, by *(4.5)*,

$$\cos(\theta+\psi)u_x+\sin(\theta+\psi)u_y = \cos(\theta+\psi)\left[\cos\theta\, u_r-\frac{\sin\theta}{r}\,u_\theta\right]$$

$$+\sin(\theta+\psi)\left[\sin\theta\, u_r+\frac{\cos\theta}{r}\,u_\theta\right]$$

$$= u_r\left[\cos(\theta+\psi)\cos\theta+\sin(\theta+\psi)\sin\theta\right]$$

$$+\frac{1}{r}\,u_\theta\left[\sin(\theta+\psi)\cos\theta-\cos(\theta+\psi)\sin\theta\right]$$

$$= \cos\psi\, u_r+\frac{\sin\psi}{r}\,u_\theta.$$

ADDITIONAL EXAMPLES

4.7. If $\phi(x, y)$ is homogeneous of degree two in x, y, show that ϕ_x and ϕ_y are homogeneous of degree one and that the second order partial derivatives of ϕ are homogeneous of degree zero. Show also that ϕ is a solution of the equation

$$x^3\frac{\partial^3\phi}{\partial x^3}+y^3\frac{\partial^3\phi}{\partial y^3} = 0.$$

4.8. If $u = F(t)$, where $t^2 = x^2+y^2+z^2$, is a solution of Laplace's equation $\partial^2 V/\partial x^2+\partial^2 V/\partial y^2+\partial^2 V/\partial z^2 = 0$ and if v is a solution of this equation which is homogeneous of degree zero in x, y, z, show that uv is also a solution of the equation.

4.9. Find the equation of the tangent plane at the point $(2, -2, 1)$ on the surface $z^3+xy+3 = 0$.

4.10. Tangent planes are drawn to the sphere $x^2+y^2+z^2 = 1$ through the point $(1, 1, 1)$. Show that the perpendiculars to them from the origin generate the cone $yz+zx+xy = 0$.

4.11. Find the derivative of the function $x^2+y^2+z^2$ in the direction of the outward normal to the surface $x^4+2y^4+3z^4 = 6$ at the point $(1, -1, 1)$.

SOLUTIONS

4.9. $2x - 2y - 3z - 5 = 0$.   **4.11.** $12/\sqrt{14}$.

5. Differentials: Jacobians

(*i*) If u is a function of the independent variables x and y, we introduce two new independent variables dx and dy, the *differentials* of x and y respectively, and we define the differential du of u, the dependent variable, by means of the relation

$$du = \frac{\partial u}{\partial x} dx + \frac{\partial u}{\partial y} dy. \tag{5.1}$$

du is sometimes called an *exact* or *total* differential and is of the form $P(x, y)dx + Q(x, y)dy$. It can be proved that a necessary and sufficient condition for $Pdx + Qdy$ to be an exact differential is

$$\frac{\partial P}{\partial y} = \frac{\partial Q}{\partial x}. \tag{5.2}$$

(*ii*) If u, v are functions of x, y, such that they define x, y as functions of u, v (in which case u, v are said to be *functionally independent*), we have

$$du = u_x dx + u_y dy, \qquad dv = v_x dx + v_y dy,$$
$$dx = x_u du + x_v dv, \qquad dy = y_u du + y_v dv.$$

On solving the first two of these equations for dx and dy, and comparing with the second two, we obtain

$$x_u = \frac{1}{J} v_y, \; x_v = -\frac{1}{J} u_y, \; y_u = -\frac{1}{J} v_x, \; y_v = \frac{1}{J} u_x, \tag{5.3}$$

where $J = u_x v_y - v_x u_y$, i.e.

$$J = \begin{vmatrix} u_x & u_y \\ v_x & v_y \end{vmatrix}.$$

We write J as $\dfrac{\partial(u, v)}{\partial(x, y)}$, the *Jacobian*, or *functional determinant* of u, v with respect to x, y. It can be shown that a necessary and sufficient condition that u, v are functionally independent is that $J \neq 0$. If a relation $F(u, v) = 0$ exists between u and v, they are *functionally dependent* (or are dependent functions) and in that case $J = 0$. It can be shown that $\partial(^x, y)/\partial u, v) = 1/\partial(u, v)/\partial(^x, y)$.

(*iii*) We extend the idea of differentials to functions of any number of variables; thus, if $u = u(x_1, x_2, \ldots, x_n)$, we define du by the relation

$$du = \frac{\partial u}{\partial x_1} dx_1 + \frac{\partial u}{\partial x_2} dx_2 + \ldots + \frac{\partial u}{\partial x_n} dx_n. \tag{5.4}$$

(*iv*) By means of Taylor's theorem (*4.1*) we see that, if Δx and Δy are numerically small, $f(x + \Delta x, y + \Delta y) - f(x, y)$ is approximately equal to

$$\frac{\partial f}{\partial x} \Delta x + \frac{\partial f}{\partial y} \Delta y.$$

If the differentials dx and dy are taken as small increments in x and y, an approximation to the increment in f is given by df, the differential of f. We shall apply this idea in examples on small errors in observations.

WORKED EXAMPLES

5.1. *Show that*

$$\frac{2x}{x^2 - y^2} dx - \frac{x^2 + y^2}{y(x^2 - y^2)} dy$$

is an exact differential and find the function u of which it is the differential.

Here

$$\frac{\partial}{\partial y} \left(\frac{2x}{x^2 - y^2} \right) = \frac{4xy}{(x^2 - y^2)^2},$$

$$\frac{\partial}{\partial x} \left(\frac{-x^2 - y^2}{y(x^2 - y^2)} \right) = \frac{-2x(x^2 - y^2) + 2x(x^2 + y^2)}{y(x^2 - y^2)^2} = \frac{4xy}{(x^2 - y^2)^2}.$$

Hence by (*5.2*) the given expression is an exact differential, i.e., there exists a function u such that

$$\frac{\partial u}{\partial x} = \frac{2x}{x^2-y^2}, \quad \frac{\partial u}{\partial y} = \frac{-(x^2+y^2)}{y(x^2-y^2)}.$$

On integrating the first of these relations with respect to x, we get

$$u = \log(x^2-y^2) + v(y),$$

where $v(y)$ is a function of y, still to be determined.

Now $$\frac{\partial u}{\partial y} = \frac{-2y}{x^2-y^2} + v'(y) = \frac{-(x^2+y^2)}{y(x^2-y^2)},$$

whence

$$v'(y) = \frac{2y^2-x^2-y^2}{y(x^2-y^2)} = -\frac{1}{y}.$$

(Note that we have a check on our working here, in that $v'(y)$ ought to be a function independent of x.)

Since $v'(y) = -\dfrac{1}{y}$, $v(y) = -\log y$ and hence

$$u = \log\left(\frac{x^2-y^2}{y}\right).$$

5.2. *Show that, if*

$$E = \frac{y}{x}\,dx + \frac{x-1}{y+1}\,dy,$$

there exists a function μ of $x+y$ alone such that μE is an exact differential and find the function u of which μE is the differential.

For $\mu E = \mu(y/x)\,dx + \mu\{(x-1)/(y+1)\}\,dy$ to be exact we require

$$\frac{\partial}{\partial y}\left(\mu\,\frac{y}{x}\right) = \frac{\partial}{\partial x}\left(\mu\,\frac{x-1}{y+1}\right),$$

i.e., $$\frac{1}{x}\mu + \frac{y}{x}\mu' = \frac{\mu}{y+1} + \frac{x-1}{y+1}\mu',$$

i.e., $$\mu\left(\frac{1}{x} - \frac{1}{y+1}\right) + \mu'\left(\frac{y}{x} - \frac{x-1}{y+1}\right) = 0,$$

i.e., $$\mu + (x+y)\mu' = 0,$$

so that we can take $\qquad \mu = (x+y)^{-1}.$

We have now

$$du = \frac{y}{x(x+y)} \, dx + \frac{x-1}{(y+1)(x+y)} \, dy,$$

so that $\qquad u = \int \left(\frac{1}{x} - \frac{1}{x+y}\right) dx = \log\left(\frac{x}{x+y}\right) + v(y), \text{ say.}$

Now

$$\frac{\partial u}{\partial y} = -\frac{1}{x+y} + v'(y) = \frac{x-1}{(y+1)(x+y)},$$

so that $v'(y) = (y+1)^{-1}.$ Hence $v(y) = \log(y+1)$ and

$$u = \log\left\{\frac{x(y+1)}{x+y}\right\}.$$

5.3. *If the relations* $u = u(x, y)$, $v = v(x, y)$, *satisfy the Cauchy-Riemann equations* $u_x = v_y$, $v_x = -u_y$ *and if they define* x, y *as functions of* u, v, *show that*

$$x_u = y_v, \quad x_v = -y_u.$$

This result follows at once by using (5.3) to transfer $u_x = v_y$, $v_x = -u_y$ to

$$Jy_v = Jx_u \quad \text{and} \quad -Jy_u = -(-Jx_v).$$

5.4. *It is given that functions* $x(u, v)$, $y(u, v)$ *satisfy the equations*

$$\frac{\partial x}{\partial u} = x \frac{\partial x}{\partial v}, \quad \frac{\partial y}{\partial u} = y \frac{\partial y}{\partial v}.$$

If u, v *are expressed as functions of* x, y, *show that*

$$(x-y) \frac{\partial^2 u}{\partial x \partial y} = \frac{\partial u}{\partial x} - \frac{\partial u}{\partial y}, \quad xy(x-y) \frac{\partial^2 v}{\partial x \partial y} = x^2 \frac{\partial v}{\partial x} - y^2 \frac{\partial v}{\partial y}.$$

Using (5.3) the given equations $x_u = xx_v$, $y_u = yy_v$ become

$$v_y = -xu_y, \quad v_x = -yu_x.$$

On differentiating the first of these with respect to x, and the second with respect to y, we get

$$v_{xy} = -u_y - xu_{xy}, \qquad v_{xy} = -u_x - yu_{xy}.$$

Hence

$$(x-y)u_{xy} = u_x - u_y.$$

Also

$$(x-y)v_{xy} = yu_y - xu_x = -\frac{y}{x}v_y + \frac{x}{y}v_x,$$

i.e.

$$xy(x-y)v_{xy} = x^2 v_x - y^2 v_y.$$

5.5. *If u and v are harmonic functions of x and y which are functionally dependent, prove that they are linearly related.*

(For the definition of harmonic functions, see **3.5.**)
Let $v = f(u)$. We have to show that $f(u)$ is linear in u.
Now

$$\frac{\partial v}{\partial x} = f'(u)\,\frac{\partial u}{\partial x}, \quad \frac{\partial^2 v}{\partial x^2} = f''(u)\left(\frac{\partial u}{\partial x}\right)^2 + f'(u)\,\frac{\partial^2 u}{\partial x^2},$$

and similarly

$$\frac{\partial^2 v}{\partial y^2} = f''(u)\left(\frac{\partial u}{\partial y}\right)^2 + f'(u)\,\frac{\partial^2 u}{\partial y^2}.$$

Since $\partial^2 u/\partial x^2 + \partial^2 u/\partial y^2 = 0$, $\partial^2 v/\partial x^2 + \partial^2 v/\partial y^2 = 0$, it follows that $f''(u) = 0$ and so $f(u) = Au + B$, where A and B are constants, i.e. u and v are linearly related.

5.6. *If u, v are functions of x, y, show that (i) $df(u) = f'(u)\,du$;* *(ii) $d(uv) = u\,dv + v\,du$;* *(iii) $d\left(\dfrac{u}{v}\right) = \dfrac{v\,du - u\,dv}{v^2}.$*

$(i)\ df(u) = \frac{\partial}{\partial x}f(u) \cdot dx + \frac{\partial}{\partial y}f(u) \cdot dy$

$$= f'(u)\left[\frac{\partial u}{\partial x}\,dx + \frac{\partial u}{\partial y}\,dy\right] = f'(u)\,du.$$

(ii) $\quad d(uv) = \dfrac{\partial}{\partial x}(uv) \cdot dx + \dfrac{\partial}{\partial y}(uv) \cdot dy$

$$= \left(u\frac{\partial v}{\partial x} + v\frac{\partial u}{\partial x} \right) dx + \left(u\frac{\partial v}{\partial y} + v\frac{\partial u}{\partial y} \right) dy$$

$$= u\left(\frac{\partial v}{\partial x}\,dx + \frac{\partial v}{\partial y}\,dy \right) + \left(\frac{\partial u}{\partial x}\,dx + \frac{\partial u}{\partial y}\,dy \right) = u\,dv + v\,du.$$

(iii) $\quad d\left(\dfrac{u}{v}\right) = \dfrac{\partial}{\partial x}\left(\dfrac{u}{v}\right) dx + \dfrac{\partial}{\partial y}\left(\dfrac{u}{v}\right) dy,$

$$= \frac{vu_x - uv_x}{v^2}\,dx + \frac{vu_y - uv_y}{v^2}\,dy,$$

$$= \frac{1}{v^2}\{v(u_x dx + u_y dy) - u(v_x dx + v_y dy)\},$$

$$= \frac{v\,du - u\,dv}{v^2}.$$

(These results clearly hold when u and v are functions of any number of variables.)

5.7. *A quantity w is calculated from measurements D, θ, R by the formula $w = D^2 \cos\theta / R^2$. Find the maximum possible percentage error in the calculated value of w, given that the measurements of D, θ, R are liable to errors of $0\cdot8\%$, 1%, $0\cdot6\%$, respectively and that θ is measured as $30°$.*

Since $\quad w = D^2 \cos\theta / R^2,$

$$\frac{\partial w}{\partial D} = \frac{2D\cos\theta}{R^2},$$

$$\frac{\partial w}{\partial \theta} = \frac{-D^2 \sin\theta}{R^2},$$

$$\frac{\partial w}{\partial R} = \frac{-2D^2 \cos\theta}{R^3} \quad \text{and so, using (5.4),}$$

$$dw = \frac{\partial w}{\partial D}\, dD + \frac{\partial w}{\partial \theta}\, d\theta + \frac{\partial w}{\partial R}\, dR$$

$$= \frac{2D \cos \theta}{R^2}\, dD - \frac{D^2 \sin \theta}{R^2}\, d\theta - \frac{2D^2 \cos \theta}{R^3}\, dR$$

$$= \frac{D^2 \cos \theta}{R^2}\left(2\,\frac{dD}{D} - \theta \tan \theta \,\frac{d\theta}{\theta} - 2\,\frac{dR}{R} \right).$$

Thus

$$\frac{dw}{w} = 2\,\frac{dD}{D} - \theta \tan \theta \,\frac{d\theta}{\theta} - 2\,\frac{dR}{R}.$$

The relative error in w is $\dfrac{dw}{w}$, so that the percentage error is $(dw/w) \times 100$, with similar definitions for the errors in D, θ and R. Hence

$$\left| \frac{dw}{w} \times 100 \right| < 2\left| \frac{dD}{d} \times 100 \right| + \left| \theta \tan \theta \right| \left| \frac{d\theta}{\theta} \times 100 \right| + 2\left| \frac{dR}{R} \times 100 \right|$$

$$= 2 \times 0.8 + \frac{\pi}{6} \times \frac{1}{\sqrt 3} \times 1 + 2 \times 0.6 = 3.1,$$

i.e. the maximum possible percentage error is 3.1.

(Note that it is necessary to express θ in radian measure, since differentiation of $\cos \theta$ is involved in the calculation.)

5.8. *The angle A of a triangle is calculated by measuring the three sides, a, b, c. If the relative errors in these measurements are α, β, γ respectively, show that the error in the calculated value of A is approximately*

$$\frac{a\alpha}{b} \operatorname{cosec} C - \beta \cot C - \gamma \cot B.$$

From the triangle formula $a^2 = b^2 + c^2 - 2bc \cos A$, we have, on taking differentials (using the results of **5.6**),

$$2a\, da = 2b\, db + 2c\, dc - 2 \cos A(c\, db + b\, dc) + 2bc \sin A\, dA,$$

$$= 2a \cos C\, db + 2a \cos B\, dc + 2bc \sin A\, dA,$$

since $b = c \cos A + a \cos C$, $c = a \cos B + b \cos A$.

Hence

$$dA = \frac{a}{bc \sin A}\, da - \frac{a \cos C}{bc \sin A}\, db - \frac{a \cos B}{bc \sin A}\, dc$$

$$= \frac{1}{b \sin C}\, da - \frac{\cos C}{b \sin C}\, db - \frac{\cos B}{c \sin B}\, dc,$$

since $c \sin A = a \sin C$, $a \sin B = b \sin A$.

Since $\alpha = da/a$, $\beta = db/b$, $\gamma = dc/c$, the error in A is approximately

$$\frac{a\alpha}{b} \operatorname{cosec} C - \beta \cot C - \gamma \cot B.$$

ADDITIONAL EXAMPLES

5.9. Show that

$$2x\left(1 + \frac{y}{x^4 + y^2}\right) dx - \frac{x^2}{x^4 + y^2}\, dy$$

is an exact differential and find the function of which it is the differential.

5.10. Verify that

$$E = \frac{y^3}{x(x + y^2)}\, dx + \frac{x - y^2}{x + y^2}\, dy$$

is not an exact differential, but that its product by a suitable function $\mu(y)$ is. Find $\mu(y)$ and the function u such that $du = E\mu(y)$.

5.11. If $x = r \cos \theta$, $y = r \sin \theta$, find $\partial(x, y)/\partial(r, \theta)$ and $\partial(r, \theta)/\partial(x, y)$ and verify that their product is unity.

5.12. If $u = x + y$, $v = x^{-1} + y^{-1}$, show that $\partial x/\partial u + \partial y/\partial u = -1$, $\partial x/\partial v + \partial y/\partial v = 0$.

5.13. A quantity w is directly proportional to x and to the square of y and inversely proportional to the cube of z. If x, y, z are measurements each liable to an error of 0.1%, show that the calculated value of w is liable to a maximum error of 0.6%.

SOLUTIONS

5.9. $x^2 - \tan^{-1}(y/x^2)$.   **5.10.** $\mu(y) = y^{-1}$, $u = \log\{xy/(x+y^2)\}$.

5.11. $\partial(x, y)/\partial(r, \theta) = r$, $\partial(r, \theta)/\partial(x, y) = r^{-1}$.

Chapter II

MAXIMA AND MINIMA

6. Functions of Two Variables

In this section we deal with maxima and minima of functions of two variables. Let $f(x, y)$ be defined in a region R of the (x, y) plane. If at least one point in R, (a, b) say, exists such that

$$f(x, y) \geq f(a, b) = m, \text{ say,}$$

for every point (x, y) in R, then m is the *absolute minimum* of $f(x, y)$ in the region R and it is attained at (a, b). There may of course be more than one point in R at which $f(x, y) = m$. An *absolute maximum* in R is defined similarly, i.e., if $f(x, y) \leq f(a', b') = M$, say, where (a', b') is in R, for all points (x, y) in R, then M is the absolute maximum of $f(x, y)$ in R. It can be shown that if $f(x, y)$ is continuous in the *closed* bounded region R, then $f(x, y)$ attains its absolute maximum and minimum in R.

Let (a, b) be an interior point of the region R of definition of $f(x, y)$. The function $f(x, y)$ is said to attain a *relative minimum* or *minimum turning value* at (a, b) if it is possible to find two positive numbers H and K such that $f(a, b) < f(a+h, b+k)$ for all values of h and k, not both zero, such that $|h| < H$, $|k| < K$. The definition of a *relative maximum* or *maximum turning value* is similar except that here $f(a, b) > f(a+h, b+k)$. In the case where $f(x, y)$ is a differentiable function it can be easily shown that a necessary condition for $f(a, b)$ to be a maximum or minimum turning value is that $f_x = f_y = 0$ at (a, b). If these conditions hold at (a, b), $f(a, b)$ is said to be a *stationary value* of $f(x, y)$. Thus for $f(a, b)$ to be a turning value it must necessarily be a stationary value. However it is easy to see that a stationary value need not be a turning value. The equation of the tangent plane at the point where $x = a, y = b$ on the surface with equation $z = f(x, y)$ is

$$(x-a)\frac{\partial f}{\partial x} + (y-b)\frac{\partial f}{\partial y} = z - f(a, b)$$

(see §4) and hence if $f(a, b)$ is a stationary value this equation reduces to $z = f(a, b)$, i.e. the tangent plane is parallel to the plane $z = 0$.

We now state a set of sufficient conditions involving second order partial derivatives of f for $f(a, b)$ to be a turning value of $f(x, y)$. We shall assume that the second order partial derivatives of f exist in a neighbourhood of (a, b) and we form the expression

$$D = f_{xx}f_{yy} - f_{xy}^2.$$

If, at (a, b), (i) $f_x = f_y = 0$, (ii) $D > 0$, then $f(a, b)$ is a turning value. If here $f_{xx} > 0$, the turning value is a minimum and if $f_{xx} < 0$, it is a maximum. If $D < 0$, the stationary value $f(a, b)$ is not a turning value. If $D = 0$, we cannot draw any conclusion as to whether or not $f(a, b)$ is a turning value, and we require further consideration (equivalent to a discussion of derivatives higher than the second).

WORKED EXAMPLES

6.1. *Discuss the maximum and minimum values of the functions f and $1/f$, where $f = \sqrt{(1 - x^2 - y^2)}$, in their regions of definition.*

The domain of f is the closed region inside and on the circle with centre the origin and radius 1. The absolute maximum of f is 1, attained at $(0, 0)$ and the absolute minimum is 0, attained at all points on the circle $x^2 + y^2 = 1$. Clearly $f(0, 0) = 1$ is also a maximum turning value, but at no point is there a minimum turning value.

For the function $1/f$ the domain is the open region defined by $x^2 + y^2 < 1$. Here we have an absolute minimum $f(0, 0) = 1$, which is also a minimum turning value. There is no absolute maximum nor any maximum turning values.

6.2. *Find the stationary values of the function*

$$f(x, y) = (x^2 + \tfrac{1}{2}xy + y^2)\, e^{x+y},$$

and examine their nature.

Here $f_x = e^{x+y}(x^2 + y^2 + \tfrac{1}{2}xy + 2x + \tfrac{1}{2}y) = 0,$

and $f_y = e^{x+y}(x^2 + y^2 + \tfrac{1}{2}xy + \tfrac{1}{2}x + 2y) = 0$

at a stationary value. On subtracting, we have $y = x$, which on substitution in either equation gives $x^2 + x = 0$, so that $x = y = 0$, or $x = y = -1$. Hence, there are two stationary values $f(0, 0) = 0$ and $f(-1, -1) = \frac{5}{2}e^{-2}$.

On differentiating further we have

$$f_{xx} = e^{x+y}(x^2 + \tfrac{1}{2}xy + y^2 + 4x + y + 2),$$
$$f_{yy} = e^{x+y}(x^2 + \tfrac{1}{2}xy + y^2 + x + 4y + 2),$$
$$f_{xy} = e^{x+y}(x^2 + \tfrac{1}{2}xy + y^2 + \tfrac{5}{2}x + \tfrac{5}{2}y + \tfrac{1}{2}).$$

Hence $D(0, 0) = (2)(2) - (\tfrac{1}{2})^2 > 0$, so that $f(0, 0)$ is a turning value. Since $f_{xx} = 2 > 0$, this turning value is a minimum.

Again, $D(-1, -1) = e^{-4}(\tfrac{1}{4} - 4) < 0$, so that $f(-1, -1)$ is not a turning value.

6.3. *A rectangular parallelepiped has sides of length x, y, z such that $x + 2y + z = 6$. Find x, y, z when the volume of the parallelepiped is greatest.*

The volume V is given by

$$V = xyz = xy(6 - x - 2y),$$

and we have

$$V_x = 6y - 2xy - 2y^2,$$
$$= 2y(3 - x - y),$$
$$V_y = 6x - x^2 - 4xy$$
$$= x(6 - x - 4y).$$

The solutions of the equations $V_x = V_y = 0$ are $(0, 0)$, $(0, 3)$, $(6, 0)$, $(2, 1)$. We need only examine the last of these, since for the first three we get zero volume.

Now $V_{xx} = -2y$, $V_{yy} = -4x$, $V_{xy} = 6 - 2x - 4y$ and so $D(2, 1) = (-2)(-8) - (-2)^2 > 0$, with $V_{xx} < 0$. Hence $(2, 1)$ gives a maximum, and the required values for the greatest volume are $x = 2$, $y = 1$, $z = 2$.

6.4. (*i*) *Find the stationary values of the function*

$$2 \log (2 + x^2 + y^2) - xy$$

and investigate their nature.

(*ii*) *Investigate the nature of the stationary value of the function* $\log(2+x^2+y^2)-xy$ *which occurs when* $x=0$, $y=0$.

(*i*) Let $$f = 2\log(2+x^2+y^2)-xy.$$

Then $$f_x = \frac{4x}{2+x^2+y^2}-y = 0,$$

and $$f_y = \frac{4y}{2+x^2+y^2}-x = 0$$

for stationary values.

Thus $\dfrac{4(x-y)}{2+x^2+y^2}+(x-y)=0$, so that $y=x$ or $6+x^2+y^2=0$, and this latter relation gives no real solutions. Taking $y=x$ and substituting in $f_x=0$, we get $x=0$, $y=0$ or $2-2x^2=0$, i.e. $x=\pm 1$. Thus the stationary values are $f(0,0)$, $f(1,1)$, $f(-1,-1)$. Now

$$f_{xx} = \frac{8-4x^2+4y^2}{(2+x^2+y^2)^2}, \quad f_{yy} = \frac{8+4x^2-4y^2}{(2+x^2+y^2)^2},$$

$$f_{xy} = \frac{-8xy-(2+x^2+y^2)^2}{(2+x^2+y^2)^2}.$$

Hence $D(0,0)=4-1>0$, $f_{xx}(0,0)>0$, so that $f(0,0)$ is a minimum turning value. Again, $D(1,1)=D(-1,-1)=\frac{1}{4}-\frac{9}{4}<0$ so that $f(1,1)$ and $f(-1,-1)$ are not turning values.

(*ii*) Let $\phi = \log(2+x^2+y^2)-xy$. It is easily verified that at $(0,0)$, $\phi_x=\phi_y=0$, so that $\phi(0,0)$ is a stationary value. We show easily that at $(0,0)$, $\phi_{xx}=\phi_{yy}=1$ and $\phi_{xy}=-1$. Hence $D(0,0)=0$ and further investigation is required to determine whether or not $\phi(0,0)$ is a turning value. Now

$$\phi(x,y) = \log 2 + \log\{1+\tfrac{1}{2}(x^2+y^2)\}-xy$$

$$= \log 2 + \tfrac{1}{2}(x^2+y^2)-\tfrac{1}{8}(x^2+y^2)^2+\ldots-xy$$

$$\simeq \log 2 + \tfrac{1}{2}(x-y)^2-\tfrac{1}{8}(x^2+y^2)^2,$$

on using the Maclaurin expansion about $X=0$ of $\log(1+X)$, where $X=\tfrac{1}{2}(x^2+y^2)$. Hence on the line $y=x$, near $(0,0)$,

$\phi < \log 2$, but away from this line there are values of (x, y) such that $\phi > \log 2$. Thus $\phi(0, 0)$ is not a turning value.

6.5. *If $z = f(x, y)$ is defined implicitly by the equation $F(x, y, z) = 0$, show that the stationary values of z occur at points (x, y, z) defined by the solutions of the simultaneous equations*

$$F = F_x = F_y = 0.$$

If $F_{xx}F_{yy} - F_{xy}^2 > 0$ and $F_{xx}/F_z > 0$ at a stationary value, show that it is a maximum. State corresponding conditions which ensure that a stationary value is (i) a minimum; (ii) neither a maximum nor a minimum. Discuss the case where

$$F = x^4 + y^4 + z^4 - 4xyz - 8.$$

Since $\partial z/\partial x = -F_x/F_z$, $\partial z/\partial y = -F_y/F_z$, stationary values of z occur where $F = F_x = F_y = 0$ provided $F_z \neq 0$. On differentiating $F_x + F_z \, \partial z/\partial x = 0$ with respect to x, we have

$$F_{xx} + F_{zx}\frac{\partial z}{\partial x} + F_{xz}\frac{\partial z}{\partial x} + F_{zz}\left(\frac{\partial z}{\partial x}\right)^2 + F_z\frac{\partial^2 z}{\partial x^2} = 0.$$

Hence at a stationary value $\partial^2 z/\partial x^2 = -F_{xx}/F_z$ and similarly $\partial^2 z/\partial y^2 = -F_{yy}/F_z$. On differentiating $F_x + F_z \, \partial z/\partial x = 0$ with respect to y, we get

$$F_{xy} + F_{xz}\frac{\partial z}{\partial y} + F_{zy}\frac{\partial z}{\partial x} + F_{zz}\left(\frac{\partial z}{\partial x}\right)\left(\frac{\partial z}{\partial y}\right) + F_z\frac{\partial^2 z}{\partial x \partial y} = 0,$$

so that at a stationary value $\partial^2 z/\partial x \partial y = -F_{xy}/F_z$. Thus at a stationary value

$$\frac{\partial^2 z}{\partial x^2} \cdot \frac{\partial^2 z}{\partial y^2} - \left(\frac{\partial^2 z}{\partial x \partial y}\right)^2 = \frac{F_{xx}F_{yy} - F_{xy}^2}{F_z^2},$$

and so if $F_{xx}F_{yy} - F_{xy}^2 > 0$, the stationary value is a turning value. If, further $F_{xx}/F_z > 0$, then $\partial^2 z/\partial x^2 < 0$ and we have a maximum turning value.

(i) If $F = F_x = F_y = 0$, $F_{xx}F_{yy} - F_{xy}^2 > 0$ and $F_{xx}/F_z < 0$, we have a minimum turning value.

(ii) If $F = F_x = F_y = 0$ and $F_{xx}F_{yy} - F_{xy} < 0$, the stationary value is neither a maximum nor a minimum turning value.

Let us now take the case where $F = x^4 + y^4 + z^4 - 4xyz - 8$. Here $F_x = 4x^3 - 4yz$, $F_y = 4y^3 - 4xz$. At a stationary value we have $y = x$ or $y = -x$. If $y = x$, then $x = 0$ or $z = x^2$. If $y = x = 0$ we get $(0, 0, \pm 2^{3/4})$. If $z = x^2 = y^2$, we get $(\pm 2^{1/2}, \pm 2^{1/2}, 2)$. If $y = -x$, we get $x = 0$ or $z = -x^2$. From $x = 0$ here we get, as before, $(0, 0, \pm 2^{3/4})$ and from $z = -x^2 = -y^2$ we get $(\mp 2^{1/2}, \pm 2^{1/2}, -2)$.

We now examine the nature of each of these stationary values. We have $F_{xx} = 12x^2$, $F_{yy} = 12y^2$, $F_{xy} = -4z$, $F_z = 4z^3 - 4xy$.

(a) At $(0, 0, \pm 2^{3/4})$, $F_{xx}F_{yy} - F_{xy}^2 < 0$. Hence at neither of these points do we have a turning value of z.

(b) At $(\pm 2^{1/2}, \pm 2^{1/2}, 2)$, $F_{xx}F_{yy} - F_{xy}^2 > 0$. Hence at both these points we have turning values of z. Now $F_{xx}/F_z > 0$, so that both give maximum turning values of z.

(c) At $(\mp 2^{1/2}, \pm 2^{1/2}, -2)$, $F_{xx}F_{yy} - F_{xy}^2 > 0$ and $F_{xx}/F_z < 0$, so that both these points give minimum turning values of z.

6.6. *Find the stationary values of the function*

$$f(x, y) = ay^2(x^2 + y^2) + 3x^4 + 4bx^3,$$

where a, b are non-zero constants. Discuss the nature of these stationary values.

Deduce the nature of the double point on the curve

$$ay^2(x^2 + y^2) + 3x^4 + 4bx^3 + b^4 = 0.$$

We have

$$f_x = 2x(ay^2 + 6x^2 + 6bx), \quad f_y = 2ay(x^2 + 2y^2),$$

and thus the equations $f_x = f_y = 0$ give $x = 0$, $y = 0$ and $x = -b$, $y = 0$. Hence the stationary values are $f(0, 0) = 0$, $f(-b, 0) = -b^4$. Again,

$$f_{xx} = 2ay^2 + 36x^2 + 24bx, \quad f_{yy} = 2ax^2 + 12ay^2,$$

$$f_{xy} = 4axy,$$

and so $D(0, 0) = 0$, $D(-b, 0) = 24ab^4$.

Since $D(0, 0) = 0$, we require some further investigation. Now $f(x, 0) = 3x^4 + 4bx^3$ and hence when $|x|$ is small, $f(x, 0)$ has the same or opposite sign to b, depending on whether x is positive or negative. Thus $f(0, 0)$ is not a turning value.

Since $D(-b, 0) = 24ab^4$, it follows that if $a > 0$, $f(-b, 0)$ is a turning value, a minimum in fact, since $f_{xx} > 0$, but that if $a < 0$, $f(-b, 0)$ is not a turning value.

We now turn to the curve

$$F(x, y) = ay^2(x^2+y^2)+3x^4+4bx^3+b^4 = 0$$
$$= f(x, y)+b^4 = 0.$$

Let (h, k) be a point on the curve. Then, using Taylor's theorem (see §4), we can write the equation of the curve as

$$F_x(x-h)+F_y(y-k)+\tfrac{1}{2}\{F_{xx}(x-h)^2+2F_{xy}(x-h)(y-k)$$
$$+F_{yy}(y-k)^2\}$$

$+$ terms of higher degree in $x-h$ and $y-k = 0$.

If at (h, k), F_x and F_y are not both zero, there is a unique tangent to the curve at (h, k) and its equation is $(x-h)F_x+(y-k)F_y = 0$.

If $F_x = F_y = 0$ and F_{xx}, F_{xy}, F_{yy} are not all zero at (h, k), then (h, k) is a *double point* on the curve and the equation of the tangents to the two branches of the curve through (h, k) is given by the line-pair

$$F_{xx}(x-h)^2+2F_{xy}(x-h)(y-k)+F_{yy}(y-k)^2 = 0.$$

If $F_{xx}F_{yy}-F_{xy}^2 < 0$, the tangents are real lines and the point is a node; if $F_{xx}F_{yy}-F_{xy}^2 > 0$, the tangents are not real and the point (h, k) is an *isolated point* on the curve.

We have already seen that $F_x = F_y = 0$ at $(0, 0)$, $(-b, 0)$. However $(0, 0)$ does not satisfy $F = 0$, while $(-b, 0)$ does. Also when $a < 0$, $F_{xx}F_{yy}-F_{xy}^2 < 0$ and the point $(-b, 0)$ is a node; when $a > 0$, $F_{xx}F_{yy}-F_{xy}^2 > 0$ and we have an isolated point. In the latter case F has a minimum turning value at $(-b, 0)$ at which $F = 0$.

6.7. *Let $f(x, y) = x^2\sqrt{(3-2x^2-2y^2)}$ and let S be the set of points (x, y) where f has real values. Find (where they exist) the sets of points (x, y) at which (i) f has a stationary value; (ii) f has a maximum turning value; (iii) f has a minimum turning value; (iv) f attains its absolute maximum in S; (v) f attains its absolute minimum in S.*

S is the closed region inside and on the circle with centre the origin and radius $\sqrt{(3/2)}$. We have

$$f_x = 2x\sqrt{(3-2x^2-2y^2)} - \frac{2x^3}{\sqrt{(3-2x^2-2y^2)}}$$

$$= \frac{2x(3-3x^2-2y^2)}{\sqrt{(3-2x^2-2y^2)}},$$

$$f_y = \frac{-2x^2y}{\sqrt{(3-2x^2-2y^2)}}.$$

Hence (*i*) $f_x = f_y = 0$ at all points in S where $x = 0$ and at the points $(\pm 1, 0)$, i.e., at these points we have stationary values. Points on the line $x = 0$ do not give turning values, since at all these points $f = 0$. At $(\pm 1, 0)$ $f_{xx} = -12$, $f_{yy} = -2$, $f_{xy} = 0$. Hence $D(\pm 1, 0) > 0$, while $f_{xx} < 0$, so that $f(\pm 1, 0)$ are maximum turning values. Thus

(*ii*) f has maximum turning values at $(\pm 1, 0)$.

(*iii*) There are no points in S at which f has a minimum turning value.

(*iv*) Since $f = 0$ on the boundary of the closed region S and $f > 0$ at all interior points of S, it follows that f attains its absolute maximum at $(\pm 1, 0)$, where it had maximum turning values.

(*v*) f attains its absolute minimum, zero, at points on the line $x = 0$ and on the circle $2x^2 + 2y^2 = 3$.

ADDITIONAL EXAMPLES

6.8. Discuss the stationary values of the following functions $f(x, y)$:

    (*i*) $y(x-3)^2 - (x-y)(x+y-6)$;
   (*ii*) $x^4 + y^4 + 6x^2y^2 + 8x^3$;
  (*iii*) $x^4 - y^4 - a^2(x^2-y^2)$ $(a \neq 0)$;
  (*iv*) $xy(x-2)(y^2-4)$;
   (*v*) $5\tan^{-1}(x^2+y^2) - 2xy$;
  (*vi*) $(x^2-y^2)(x^2+y^2+2x+1)$;
 (*vii*) $e^{-(1/10)(x^2+y^2)}(x+2y)$;
(*viii*) $xy(x^2+y^2+2x+2y+2)$;
  (*ix*) $x^3 - x^2y - x^2 + y^2$;
   (*x*) $x^4 + y^4 - x^2 + 2xy - y^2$.

6.9. Show that for all values of the constant a the function $x^3 - 3xy^2 + (a+3)y^2 - 2ay$ has a stationary value when $x = y = 1$ and investigate its nature (*i*) when $a = 4$; (*ii*) when $a = 3$.

6.10. Examine for double points the curves

$$y^2 = x + \frac{k}{x-4},$$

when k has the values $3, 4, -5$.

Discuss the stationary values of the function

$$(x-4)(y^2-x).$$

6.11. Discuss the stationary values of $z(x, y)$, which is defined implicitly by the relation $x^3 + y^3 + z^3 + 3xyz - 1 = 0$.

6.12. Find the shortest distance from the origin to points on the surface with equation $z^2 = x + y + 1$.

SOLUTIONS

6.8. (*i*) $f(3, 3)$, $f(5, 1)$, $f(1, 1)$ are stationary values, the first a minimum, the other two are not turning values.

(*ii*) $f(0, 0)$ is a stationary value but not a turning value, $f(-6, 0)$ is a minimum.

(*iii*) $f(0, 0)$ is a stationary value but not a turning value, $f(\pm a/\sqrt{2}, 0)$ are minima, $f(0, \pm a/\sqrt{2})$ are maxima, $f(a/\sqrt{2}, \pm a/\sqrt{2})$, $f(-a/\sqrt{2}, \pm a/\sqrt{2})$ are stationary values but not turning values.

(*iv*) $f(0, 0)$, $f(2, 0)$, $f(0 \pm 2)$, $f(2 \pm 2)$ are stationary values but not turning values, $f(1, 2/\sqrt{3})$ is a maximum, $f(1, -2/\sqrt{3})$ is a minimum.

(*v*) $f(0, 0)$ is a minimum, $f(\pm 1, \pm 1)$ are stationary values but not turning values.

(*vi*) $f(0, 0)$, $f(-\frac{1}{2}, 0)$ are stationary values but not turning values, $f(-1, 0)$ is a minimum.

(*vii*) $f(1, 2)$ is a maximum, $f(-1, -2)$ is a minimum.

(*viii*) $f(0, 0)$ is a stationary value but not a turning value, $f(-1, -1)$ is a minimum, $f(-\frac{1}{2}, -\frac{1}{2})$ is a stationary value but not a turning value.

(ix) $f(0, 0)$, $f(2, 2)$ are stationary values but not turning values, $f(1, \frac{1}{2})$ is a minimum.

(x) $f(0, 0)$ is a stationary value but not a turning value, $f(\pm 1, \mp 1)$ are minima.

6.9. (i) $f(1, 1)$ is a minimum. (ii) $f(1, 1)$ is not a turning value.

6.10. $k = 3, -5$, no double points, $k = 4$, isolated point at $(2, 0)$; $f(2, 0)$ is a maximum (corresponding to the case where $k = 4$), $f(4, \pm 2)$ are stationary values but not turning values.

6.11. $z(0, 0) = 1$ is a stationary value but not a turning value; $z(-2^{-1/3}, -2^{-1/3}) = 2^{-1/3}$ is a minimum.

6.12. $1/\sqrt{2}$.

7. Functions of More than Two Variables

We now deal with maxima and minima of $f(x_1, \ldots, x_n)$, where n is any positive integer, in a region R of $(x_1, \ldots, x_n)$ space. The definitions of absolute maxima and minima in R are exactly similar to those for $f(x, y)$ given in §6 and similarly for the definition of turning values at interior points of R. If the first order partial derivatives of f exist and $(a_1, \ldots, a_n)$ is an interior point of R, we say that $f(a_1, \ldots, a_n)$ is a stationary value of f if $\partial f/\partial x_1, \ldots, \partial f/\partial x_n$ are zero at $(a_1, \ldots, a_n)$. As before it is necessary but not sufficient for $f(a_1, \ldots, a_n)$ to be a turning value that it should be a stationary value.

When the second order partial derivatives exist, we can state sufficient conditions in terms of these derivatives that a stationary value should be a turning value.

Let f_{ij} denote the value of $\partial^2 f/\partial x_i \partial x_j$ at a point giving a stationary value. Then f is a minimum turning value if the following n inequalities hold

$$f_{11} > 0, \quad \begin{vmatrix} f_{11} & f_{12} \\ f_{21} & f_{22} \end{vmatrix} > 0, \ldots, \quad \begin{vmatrix} f_{11} \cdots f_{1n} \\ \cdots\cdots\cdots \\ f_{n1} \cdots f_{nn} \end{vmatrix} > 0,$$

and it is a maximum turning value if

$$f_{11} < 0, \quad \begin{vmatrix} f_{11} & f_{12} \\ f_{21} & f_{22} \end{vmatrix} > 0, \ldots, (-1)^n \begin{vmatrix} f_{11} & \cdots & f_{1n} \\ \cdots & \cdots & \cdots \\ f_{n1} & \cdots & f_{nn} \end{vmatrix} > 0$$

However, as will be seen in the following problems, it is usually better to consider from first principles the sign of

$$f(a_1+h_1, \ldots, a_n+h_n) - f(a_1, \ldots, a_n)$$

for small values of $|h_1|, \ldots, |h_n|$ rather than to evaluate these determinants.

WORKED EXAMPLES

7.1. *Examine for turning values the function*

$$f(x, y, z) = (y+z)(x^2+y^2+z^2-6).$$

For a stationary value we require

$$f_x = 2x(y+z) = 0,$$
$$f_y = 2y(y+z)+x^2+y^2+z^2-6 = 0,$$
$$f_z = 2z(y+z)+x^2+y^2+z^2-6 = 0.$$

From the second and third of these relations we obtain $(y-z)(y+z) = 0$, and, from the first, $x = 0$ or $y+z = 0$. Hence stationary values occur at the points $(0, \pm1, \pm1)$ and at all points where the plane $y+z = 0$ meets the cylinder $x^2+2y^2 = 6$. The only possible turning values are $f(0, \pm1, \pm1)$.

We now examine $f(0, 1, 1)$. Let $x' = x, y' = y-1, z' = z-1$ and consider

$$f(x, y, z) - f(0, 1, 1)$$
$$= (y'+z'+2)[x'^2+(y'+1)^2+(z'+1)^2-6]+8$$
$$= (y'+z'+2)(x'^2+y'^2+z'^2+2y'+2z'-4)+8$$
$$= 2x'^2+4y'^2+4z'^2+4y'z'+\text{third degree terms}$$
$$= 2x'^2+4(y'+\tfrac{1}{2}z')^2+3z'^2+\text{third degree terms}.$$

Now the quadratic form $2x'^2+4(y'+\tfrac{1}{2}z')^2+3z'^2$ is *positive definite*, i.e., it is always positive for values of x', y', z' which are not all zero, and it follows that $f(0, 1, 1)$ is a minimum turning value. From symmetry we see that $f(0, -1, -1)$ is also a minimum turning value.

7.2. *Examine for turning values the function*

$$f(x, y, z) = x^4 + y^4 + z^4 - 4xyz.$$

For a stationary value we require

$$f_x = 4x^3 - 4yz = 0, \; f_y = 4y^3 - 4zx = 0, \; f_z = 4z^3 - 4xy = 0.$$

Hence $x^4 = y^4 = z^4$; if $x = y = z$, we get $4x^2(x-1) = 0$ giving $(0, 0, 0)$, $(1, 1, 1)$; if $x = y = -z$ or $x = -y = z$, we get $4x^2(x+1) = 0$ giving $(-1, -1, 1)$, $(-1, 1, -1)$; if $x = -y = -z$, we get $4x^2(x-1) = 0$, giving $(1, -1, -1)$. and so the stationary values are $f(0, 0, 0)$, $f(1, 1, 1)$, $f(1, -1, -1)$, $f(-1, 1, -1)$, $f(-1, -1, 1)$.

$f(0, 0, 0) = 0$ is not a turning value, since near $(0, 0, 0)$ the dominant term in f is $-4xyz$ and thus f takes both positive and negative values.

To examine $f(1, 1, 1)$, let $x = 1+h$, $y = 1+k$, $z = 1+l$ and consider

$$f(1+h, 1+k, 1+l) - f(1, 1, 1) = 6(h^2 + k^2 + l^2) - 4(hk + kl + lh)$$
$$+ \text{terms of higher degree.}$$

Now $6(h^2 + k^2 + l^2) - 4(hk + kl + lh)$
$$= 2h^2 + 2k^2 + 2l^2 + 2(h-k)^2 + 2(k-l)^2 + 2(l-h)^2,$$

which is always greater than zero unless $h = k = l = 0$. Thus $f(1, 1, 1)$ is a minimum turning value.

To examine $f(-1, -1, 1)$, let $x = -1+h$, $y = -1+k$, $z = 1+l$ and consider

$$f(-1+h, -1+k, 1+l) - f(-1, -1, 1)$$
$$= 6(h^2 + k^2 + l^2) + 4(-hk + kl + lh) + \text{terms of higher degree.}$$

Now $3(h^2 + k^2 + l^2) + 2(-hk + kl + lh)$

$$= 3\left[\left(h - \frac{k}{3} + \frac{l}{3}\right)^2\right] + \frac{8}{3}k^2 + \frac{8}{3}l^2 + \frac{8}{3}kl$$

$$= 3\left[\left(h - \frac{k}{3} + \frac{l}{3}\right)^2\right] + \frac{8}{3}(k + \tfrac{1}{2}l)^2 + 2l^2,$$

which is a positive definite quadratic form and hence $f(-1, -1, 1)$ is a minimum turning value.

Similarly $f(1, -1, -1)$ and $f(-1, 1, -1)$ are minima.

We could have calculated the values of the determinants, mentioned above, with elements the second order partial derivatives and we would have found that they were all positive, showing that the stationary values [except for $f(0, 0, 0)$] are minimum turning values.

7.3. Let $f(x, y, z) = x^2\sqrt{(6-x^2-y^2-z^2)}$ and let S be the set of points where f has real values. Find (where they exist) the sets of points (x, y, z) at which (i) f has a stationary value; (ii) f has a maximum turning value; (iii) f has a minimum turning value; (iv) f attains its absolute maximum in S; (v) f attains its absolute minimum in S.

We have
$$f_x = 2x\sqrt{(6-x^2-y^2-z^2)} - \frac{x^3}{\sqrt{(6-x^2-y^2-z^2)}}$$
$$= \frac{x(12-3x^2-2y^2-2z^2)}{\sqrt{(6-x^2-y^2-z^2)}} = 0,$$
$$f_y = \frac{-x^2y}{\sqrt{(6-x^2-y^2-z^2)}} = 0,$$
$$f_z = \frac{-x^2z}{\sqrt{(6-x^2-y^2-z^2)}} = 0$$

at a stationary value. S is the set of points where $x^2+y^2+z^2 \leqslant 6$.

(i) The stationary values occur at the points $(\pm 2, 0, 0)$ and at all points in plane $x = 0$ where $x^2+y^2+z^2 \leqslant 6$.

(ii) The only possible points at which f can have turning values are $(\pm 2, 0, 0)$. We now examine the nature of the stationary value at $(2, 0, 0)$. Let $x = 2+h, y = k, z = l$ and we have

$$f(x, y, z) - f(2, 0, 0)$$
$$= (2+h)^2\sqrt{(2-4h-h^2-k^2-l^2)} - 4\sqrt{2}$$
$$= (4+4h+h^2)\sqrt{2}\left(1-2h-\frac{h^2+k^2+l^2}{2}\right)^{1/2} - 4\sqrt{2}$$
$$= 4\sqrt{2}\left(1+h+\frac{h^2}{4}\right)\left(1-h-\frac{h^2+k^2+l^2}{4}-\tfrac{1}{2}h^2 + \ldots\right) - 4\sqrt{2}$$
$$= 4\sqrt{2}\left(-\frac{3}{2}h^2-\frac{k^2}{4}-\frac{l^2}{4}\right) + \ldots.$$

Hence $f(2, 0, 0)$ is a maximum turning value and similarly for $f(-2, 0, 0)$.

(*iii*) Clearly there are no minimum turning values.

(*iv*) Since $f = 0$ on $x = 0$ and on $x^2 + y^2 + z^2 = 6$ and elsewhere in S it is positive, and since S is a closed region, f must attain its absolute maximum at one or more interior points of S. These must also yield turning values. Hence the absolute maximum of f in S is $4\sqrt{2}$ attained at $(\pm 2, 0, 0)$.

(*v*) The absolute minimum of S is 0, attained at all points in S where $x = 0$ or $x^2 + y^2 + z^2 = 6$.

ADDITIONAL EXAMPLES

7.4. Show that the function

$$\tan^{-1}(x^2 + y^2 + z^2) - (yz + zx + xy)$$

has a stationary value only when $x = y = z = 0$. Find the nature of this stationary value.

7.5. Find the five points at which the function

$$2xyz + x^2 + y^2 + z^2$$

has a stationary value. Show that at exactly one of these points the function has a turning value and determine its nature.

7.6. Verify that the function

$$f(x, y, z) = \log(2k + x^2 + y^2 + z^2) - yz - zx - xy,$$

where $k > 0$ has a stationary value at $x = y = z = 0$. Show that if $k < \tfrac{1}{2}$ this stationary value is a minimum turning value and that if $k \geqslant \tfrac{1}{2}$ it is not a turning value.

SOLUTIONS

7.4. This stationary value is not a turning value.

7.5. Stationary values at $(0, 0, 0)$, $(-1, -1, -1)$, $(-1, 1, 1)$, $(1, -1, 1)$, $(1, 1, -1)$. The only turning value is at $(0, 0, 0)$ and it is a minimum.

8. Lagrange's Method of Undetermined Multipliers

Lagrange's method of undetermined multipliers is a means of finding stationary values of functions whose arguments are connected by one or more relations, which we shall call *equations of condition*. The general case is as follows: let $u = f(x_1, \ldots, x_n)$, where $\phi_r(x_1, \ldots, x_n) = 0$, $r = 1, \ldots, m$ $(m < n)$. Then it can be shown that at a stationary value of u we have

$$\frac{\partial f}{\partial x_i} + \sum_{r=1}^{m} \lambda_r \frac{\partial \phi_r}{\partial x_i} = 0, \quad i = 1, \ldots, n, \qquad (8.1)$$

where λ_r are the undetermined multipliers. Equations (8.1) are called Lagrange's equations. These equations together with the m equations of condition enable us to determine the values of $x_1, \ldots, x_n$ (and $\lambda_1, \ldots, \lambda_m$) at stationary values.

To determine the nature of a stationary value obtained by Lagrange's method, we have to investigate further. Often we can use geometrical considerations, as will be shown in some of the following examples.

WORKED EXAMPLES

8.1. *Find the stationary values of the function* $x^2 + y^2 - z^2$, *where the variables are connected by the equation* $yz + zx + xy + 1 = 0$. *Discuss their nature.*

Let $u = x^2 + y^2 - z^2$. The Lagrange equations are

$$2x + \lambda(y+z) = 0, \qquad (1)$$

$$2y + \lambda(z+x) = 0, \qquad (2)$$

$$-2z + \lambda(x+y) = 0. \qquad (3)$$

$(1) \times x + (2) \times y + (3) \times z$ gives $2u - 2\lambda = 0$,

i.e., $\lambda = u$. We now eliminate x, y, z from the equations to obtain

$$\begin{vmatrix} 2 & u & u \\ u & 2 & u \\ u & u & -2 \end{vmatrix} = 0,$$

i.e., $u^3 - u^2 - 4 = 0$, i.e., $(u-2)(u^2+u+2) = 0$. Hence 2 is the only stationary value of u. When $u = 2$, the Lagrange equations become $x+y+z = 0$, $x+y-z = 0$, whence $y = -x$, $z = 0$. Substituting these in the equation of condition we see that the stationary value 2 is attained at $(1, -1, 0)$ and $(-1, 1, 0)$.

To examine the nature of $u(1, -1, 0)$ we consider

$$u(1+h, -1+k, l) - u(1, -1, 0) = 2(h-k)+h^2+k^2-l^2.$$

We have to remember that $(1+h, -1+k, l)$ must satisfy the equation of condition,

i.e., $$(1+h)(-1+k)+l(h+k)+1 = 0,$$

i.e., $$-h+k+hk+lh+lk = 0. \tag{4}$$

Thus,

$$u(1+h, -1+k, l) - u(1, -1, 0) = h^2+k^2-l^2+2hk+2lh+2lk.$$

Now from (4) we see that

$$k = h + \text{second degree terms},$$

so that

$$u(1+h, -1+k, l) - u(1, -1, 0) = 4h^2 - l^2 + 4lh + \text{terms of higher degree}.$$

The quadratic form $4h^2 - l^2 + 4lh$ is indefinite, that is, it can take both positive and negative values. Hence $u(1, -1, 0)$ is not a turning value. From symmetry we see that neither is $u(-1, 1, 0)$.

8.2. *Find the lengths and equations of the axes of the section of the quadric $4yz + 5zx - 5xy = 8$ by the plane $x+y-z = 0$.*

This problem can be thought of in terms of the fact that the axes are maximum or minimum values of the diameters of the section of the quadric. If (x, y, z) is the extremity of a diameter, then the length of the diameter is $2\sqrt{(x^2+y^2+z^2)}$, where, of course, $x+y-z = 0$ and $4yz+5zx-5xy = 8$. It will be equivalent, and simpler, if we study the maxima and minima of $u = x^2+y^2+z^2$ subject to the equations of condition $x+y-z = 0$ and $4yz+5zx-5xy = 8$.

The Lagrange equations are

$$2x+\lambda(-5y+5z)+\mu = 0, \qquad (1)$$

$$2y+\lambda(-5x+4z)+\mu = 0, \qquad (2)$$

$$2z+\lambda(\;\;5x+4y)-\mu = 0. \qquad (3)$$

Now $(1) \times x + (2) \times y + (3) \times z$ gives $2u+16\lambda = 0$,
i.e., $\lambda = -\frac{1}{8}u$. Also $(2)+(3)$ gives $2(y+z)+4\lambda(y+z) = 0$,
i.e., $\lambda = -\frac{1}{2}$ or $y+z = 0$.

If $\lambda = -\frac{1}{2}$, then $u = 4$ and the length of the corresponding axis
is $2\sqrt{4} = 4$. With $\lambda = -\frac{1}{2}$, $(1)-(2)$ gives $-x+y-z = 0$. Also we
have the equation of condition $x+y-z = 0$ and hence $x = 0$, $y = z$ is
the equation of this axis.

For the other axis we have $y+z = 0$, and this, together with
$x+y-z = 0$, gives

$$\frac{x}{2} = \frac{y}{-1} = \frac{z}{1} \quad (= k).$$

This is the equation of the axis. It meets the quadric where
$k^2(-4+10+10) = 8$, i.e. $k = \pm 1/\sqrt{2}$. Hence $u = (4+1+1)k^2 = 3$,
so that the length of this axis is $2\sqrt{3}$.

From geometrical considerations it is clear that one of the stationary values is a maximum, corresponding to the major axis of the
section (an ellipse) while the other is a minimum, corresponding to
the minor axis.

8.3. *Examine for stationary values the function*

$$u = x^2+\tfrac{1}{8}y^2+\tfrac{1}{27}z^2, \text{ where } x^{1/2}+y^{1/2}+z^{1/2} = 6a^{1/2}$$

$(a > 0)$. *Investigate the nature of the stationary value. Show that*
$6a^2 \leqslant u \leqslant 1296a^2$, *and explain why the greatest value is not given by
the calculus method.*

Clearly we are dealing only with non-negative values of x, y, z.
The Lagrange equations are

$$2x+\tfrac{1}{2}\lambda x^{-1/2} = 0, \quad \tfrac{1}{4}y+\tfrac{1}{2}\lambda y^{-1/2} = 0, \quad \tfrac{2}{7}z+\tfrac{1}{2}\lambda z^{-1/2} = 0,$$

so that $2x^{3/2} = \tfrac{1}{4}y^{3/2} = \tfrac{2}{7}z^{3/2}$, whence $4x = y = \tfrac{4}{9}z$. Thus from
$x^{1/2}+y^{1/2}+z^{1/2} = 6a^{1/2}$ we obtain

$$x^{1/2}(1+2+3) = 6a^{1/2}, \text{ i.e., } x = a, \ y = 4a, \ z = 9a.$$

Hence $u(a, 4a, 9a) = 6a^2$ is the only stationary value.

We shall examine its nature using geometrical considerations. Let $x = \xi$, $\frac{1}{4}\sqrt{2}y = \eta$, $\frac{1}{9}\sqrt{3}z = \zeta$; then $u = \xi^2+\eta^2+\zeta^2$ where

$$\xi^{1/2}+2^{3/4}\eta^{1/2}+3^{3/4}\zeta^{1/2} = 6a^{1/2}. \tag{1}$$

Then u is the square on the distance from the origin to the point (ξ, η, ζ) on the surface (1). We are dealing with the octant where ξ, η, ζ are all non-negative and the shape of the surface in this octant is convex towards the origin. Thus the stationary value which occurs at $\xi = a$, $\eta = \sqrt{2}a$, $\zeta = \sqrt{3}a$ must be a minimum turning value. Also from the shape of the surface we see that the point on the surface which is at the greatest distance from the origin is the point $\xi = 36a$, $\eta = \zeta = 0$. At this point $u = 1296a^2$. Hence

$$6a^2 \leqslant u \leqslant 1296a^2.$$

The greatest value $1296a^2$ is not given by the Lagrange method since the partial derivatives do not exist at the point where this value is attained.

8.4. *If p is the sum of the lengths of all the edges of a solid rectangular block and if S is the total surface area, prove that $p^2 \geqslant 24S$. If $p = 30$ and $S = 36$, show that the volume V of the block is such that*

$$27/2 \leqslant V \leqslant 14.$$

Let x, y, z be the lengths of the edges; then $p = 4(x+y+z)$, $S = 2(yz+zx+xy)$. Hence

$$p^2-24S = 16(x+y+z)^2 -48(yz+zx+xy)$$

$$= 16(x^2+y^2+z^2 -yz-zx-xy)$$

$$= 8[(y-z)^2+(z-x)^2+(x-y)^2] \geqslant 0,$$

i.e. $\qquad\qquad p^2 \geqslant 24S.$

In the case where $p = 30$, $S = 36$, we have $V = xyz$, where

$$x+y+z = \tfrac{15}{2}, \tag{1}$$

$$yz+zx+xy = 18. \tag{2}$$

The Lagrange equations for this problem are

$$yz+\lambda+(y+z)\mu = 0, \tag{3}$$

$$zx+\lambda+(z+x)\mu = 0, \tag{4}$$

$$xy+\lambda+(x+y)\mu = 0. \tag{5}$$

$(4)-(5)$ gives $x(z-y)+\mu(z-y) = 0$, i.e. $z = y$ or $x = -\mu$. If we put $z = y$ in (1) and (2) we get

$$x+2y = \tfrac{15}{2}, \quad 2xy+y^2 = 18,$$

whence $y^2-5y+6 = 0$, i.e., $y = 2, 3$ and hence $x = \tfrac{7}{2}, \tfrac{3}{2}$. Thus $(\tfrac{7}{2}, 2, 2)$ and $(\tfrac{3}{2}, 3, 3)$ give stationary values and so, from symmetry, do $(2, \tfrac{7}{2}, 2)$, $(2, 2, \tfrac{7}{2})$, $(3, \tfrac{3}{2}, 3)$, $(3, 3, \tfrac{3}{2})$.

If we put $x = -\mu$ in (3) and (4) we get

$$yz+\lambda-xy-xz = 0,$$

$$zx+\lambda-zx-x^2 = 0,$$

whence at once $y = x$ or $z = x$. Thus we get again the above stationary values. The case where $x = y = z$ may be discarded since then the equations of condition give inconsistent values for x, y, z.

Summing up we see that V has a stationary value 14 at $(\tfrac{7}{2}, 2, 2)$, $(2, \tfrac{7}{2}, 2)$, $(2, 2, \tfrac{7}{2})$ and $\tfrac{27}{2}$ at $(\tfrac{3}{2}, 3, 3)$, $(3, \tfrac{3}{2}, 3)$, $(3, 3, \tfrac{3}{2})$.

Now geometrically we are discussing the values of $V (= xyz)$ at points on the ellipse which give the section of the quadric $xy+yz+zx = 18$ by the plane $x+y+z = \tfrac{15}{2}$. As we move round the ellipse V varies continuously and hence $\tfrac{27}{2}$ is the least and 14 the greatest value of V, i.e. $\tfrac{27}{2} \leqslant V \leqslant 14$.

8.5. *Find the stationary values of the function*

$$x^2+6xy+2y^2 \text{ where } 10x^2+4xy+13y^2 = 21.$$

Find the nature of these stationary values.

Deduce that the function $10x^2+4xy+13y^2$, *where* $x^2+6xy+2y^2=9$ *has only one stationary value, and find its nature.*

Let $u = x^2+6xy+2y^2$, where $10x^2+4xy+13y^2 = 21$. The Lagrange equations are

C

$$x+3y+\lambda(10x+2y) = 0,$$

$$3x+2y+\lambda(2x+13y) = 0,$$

so that

$$(x+3y)(2x+13y)-(3x+2y)(10x+2y) = 0,$$

i.e., $4x^2+xy-5y^2 = 0$, i.e., $y = x$ or $5y = -4x$.

When $y = x$, we obtain from the equation of condition $27x^2 = 21$; i.e., $x = \pm\frac{1}{3}\sqrt{7}$, $y = \pm\frac{1}{3}\sqrt{7}$. When x and y have these values, we obtain the stationary value $u = 7$. When $5y = -4x$, we obtain $x = \pm\frac{5}{6}\sqrt{2}$, $y = \mp\frac{2}{3}\sqrt{2}$, giving the stationary value $u = -\frac{7}{2}$.

We can obtain the nature of these stationary values from geometrical considerations. The curve with equation $10x^2+4xy+13y^2 = 21$ is an ellipse while the equation $x^2+6xy+2y^2 = c$ defines a set of hyperbolas as c varies. We see from the Lagrange equations that at a point (x, y) which yields a stationary value for u we have

$$-\frac{x+3y}{3x+2y} = -\frac{10x+2y}{2x+13y}.$$

Here the right-hand side is the gradient of the ellipse at the point (x, y) on it and the left-hand side is the gradient at (x, y) of the hyperbola of the set which passes through (x, y). Thus these two curves touch at (x, y). For values of c just greater than the stationary value 7 the hyperbolas do not intersect the ellipse in real points, while for c just less than 7 they do. Hence $u = 7$ is a maximum. Similar reasoning shows that $u = -7/2$ is a minimum.

In the second problem $x^2+6xy+2y^2 = 21$ is a hyperbola and we consider the set of ellipses $10x^2+4xy+13y^2 = c$.

Only one of these touches the hyperbola and hence we have only one stationary value, which is a minimum turning value. The Lagrange equations have the same solutions $y = x$ and $5y = -4x$ as above, but only one of these, viz. $y = x$, leads to real solutions. This corresponds to the minimum turning value.

8.6. *Find the volume of the rectangular parallelepiped of maximum volume which can be inscribed in the ellipsoid*

$$\frac{x^2}{a^2}+\frac{y^2}{b^2}+\frac{z^2}{c^2} = 3.$$

From considerations of geometrical symmetry the parallelepiped must have the origin as its centre point and its edges parallel to the coordinate axes. Let the lengths of the edges be $2x$, $2y$, $2z$. Then the volume is $8xyz$ where $x^2/a^2 + y^2/b^2 + z^2/c^2 = 3$.

It is easier to change the variables by letting $\xi = x/a$, $\eta = y/b$, $\zeta = z/c$. We now look for stationary values of $8abc\xi\eta\zeta$ where $\xi^2 + \eta^2 + \zeta^2 = 3$. Let $u = \xi\eta\zeta$ (ξ, η, ζ non-negative), where $\xi^2 + \eta^2 + \zeta^2 = 3$. The Lagrange equations are

$$\eta\zeta + 2\lambda\xi = 0, \quad \zeta\xi + 2\lambda\eta = 0, \quad \xi\eta + 2\lambda\zeta = 0,$$

whence $\xi = \eta = \zeta$ and the equation of condition gives $\xi = \eta = \zeta = 1$.

We now discuss the nature of this stationary value $u(1, 1, 1) = 1$. Let us consider

$$E = u(1+h, 1+k, 1+l) - u(1, 1, 1) = (1+h)(1+k)(1+l) - 1$$

$$= h+k+l+hk+kl+lh+hkl,$$

where $(1+h)^2 + (1+k)^2 + (1+l)^2 = 3$, i.e. where

$$2(h+k+l) + h^2 + k^2 + l^2 = 0;$$

i.e.

$$E = -\tfrac{1}{2}(h^2+k^2+l^2) + hk + kl + lh + hkl$$

$$= -\tfrac{1}{2}(h^2+k^2+l^2 - 2hk - 2kl - 2lh) + hkl.$$

But $l = -(h+k) +$ second degree terms, so that

$$E = -\tfrac{1}{2}[h^2+k^2+(h+k)^2 - 2hk + 2(h+k)^2] + \text{terms of higher degree.}$$

$$= -2(h^2+k^2+hk) + \text{terms of higher degree.}$$

Now $h^2+k^2+hk > 0$ for all values of h, k not both zero. Hence for small values of $|h|$, $|k|$ (not both zero), $E < 0$. Thus $u(1, 1, 1)$ is a maximum turning value.

Hence the required maximum volume is $8abc$.

ADDITIONAL EXAMPLES

8.7. The tangent plane at a variable point on the ellipsoid

$x^2/a^2 + y^2/b^2 + z^2/c^2 = 1$ meets the coordinate axes in L, M, N. Find the minimum distance of the centroid of the triangle LMN from the origin of coordinates.

8.8. The variables in the function $x^2 + 3y^2 + z^2 - 2xz$ are connected by the equation $x^2 + 4yz - 8 = 0$; find the four sets of values (x, y, z) at which the function has a stationary value. Discuss whether or not the stationary value which occurs at $(-2, 1, 1)$ is a turning value.

8.9. Show that the perpendicular distance from the origin to any tangent plane to the surface $xy^2z^3 = a^6$ is not greater than $2^{1/3}\, 3^{1/4} a$.

8.10. Show that the lengths of the axes of the section of the quadric $x^2 + y^2 + z^2 + yz + zx + xy = 1$ by the plane $x + 2y - 3z = 0$ are $2\sqrt{2}$ and $\sqrt{2}$. Find the equations of the axes.

8.11. Find the stationary values of the function $x^2 + xy + y^2$, where the variables are connected by the relation $5x^2 + 4xy + 4y^2 = 16$. Investigate the nature of these stationary values.

8.12. Find the stationary values of the function $x^2 + y^2 - z^2$ where the variables are connected by the equation $yz + zx + xy + 1 = 0$. Discuss the nature of these stationary values.

8.13. Show that the perpendicular distance from the origin to any tangent plane to the surface

$$\left(\frac{x}{a}\right)^{1/2} + \left(\frac{y}{b}\right)^{1/2} + \left(\frac{z}{c}\right)^{1/2} = 1 \quad (a, b, c > 0, x, y, z \geq 0)$$

is not greater than $(a^{-2/3} + b^{-2/3} + c^{-2/3})^{-3/2}$.

8.14. Let f, g, h, ϕ be four functions of the variables x, y, z. It is required to find the stationary values of the product fgh under the condition $\phi = 0$. Verify that fgh has a stationary value at a point where ϕ and at least two of f, g, h vanish. By considering the Lagrange equations as linear equations in gh, hf, fg, show that stationary values occur when

$$f \frac{\partial(g, h, \phi)}{\partial(x, y, z)} = g \frac{\partial(h, f, \phi)}{\partial(x, y, z)} = h \frac{\partial(f, g, \phi)}{\partial(x, y, z)}.$$

SOLUTIONS

8.7. $\frac{1}{3}(a+b+c)$.   **8.8.** Stationary values at $(\pm2\sqrt{2}, 0, \pm2\sqrt{2})$, $(\mp2, \pm1, \pm1)$, $f(-2, 1, 1)$ is not a turning value. **8.10.** Axes are $x/-5 = y/4 = z$, $x = y = z$. **8.11.** Stationary values at $(0, \pm2)$, $(\pm2, \mp1)$; the first two are maxima, the second two minima.

8.12. Stationary values at $(\pm1, \pm1, 0)$, neither is a turning value.

Chapter III

MULTIPLE INTEGRALS

9. Double Integrals

Let $f(x, y)$ be a bounded function defined inside and on the boundary of a closed bounded area K of the xy plane. Lines drawn parallel to the x and y axes divide K into small rectangles (and parts of rectangles near the boundary of K). Let (x, y) be any point in one of the small rectangles, whose sides have lengths Δx and Δy. We form the sum

$$\Sigma = \Sigma f(x, y)\, \Delta x\, \Delta y$$

over all the rectangles and then let the number of rectangles tend to infinity so that each Δx, Δy tends to zero. If when this is done, Σ tends to a limit, this limit is the *double integral* of $f(x, y)$ over the area K and is denoted by

$$\iint\limits_K f(x, y)\, dxdy.$$

K is called the *field of integration*. The double integral measures the volume of the solid figure bounded by the plane $z = 0$, the surface $z = f(x, y)$ and the cylinder through the boundary of K with generators parallel to the z-axis.

We shall assume that K is such that its boundary is met by any line parallel to the x- and y-axes in at most two points. If this is not so, we divide K into portions for which it is true and add the integrals over the separate portions.

In Fig. 6 let the equations of the curves ABC, ADC be $y = \phi_1(x)$, $y = \phi_2(x)$ respectively. By performing the summation Σ first taking all the rectangles in a column parallel to the y-axis and then over all such columns, we obtain the limit (i.e., the double integral) as a *repeated integral*.

$$\int_a^b \left\{ \int_{\phi_1(x)}^{\phi_2(x)} f(x, y)\, dy \right\} dx,$$

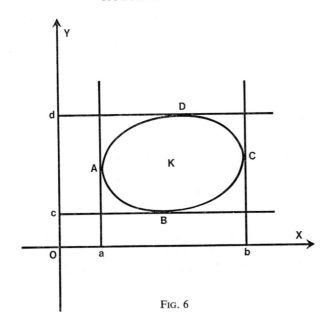

FIG. 6

written as

$$\int_a^b dx \int_{\phi_1(x)}^{\phi_2(x)} f(x, y)\,dy. \tag{9.1}$$

In the integral with respect to y, called the *inner integral*, x is held constant. We can clearly interchange the parts played by x and y in this process to obtain a second repeated integral form for the double integral:

$$\int_c^d dy \int_{\psi_1(y)}^{\psi_2(y)} f(x, y)\,dx, \tag{9.2}$$

where $x = \psi_1(y)$, $x = \psi_2(y)$ are the equations of the curves DAB, DCB respectively. When we change from (9.1) to (9.2) or (9.2) to (9.1), we are said to have *changed the order of integration*. It often happens that one of the repeated integrals is much easier to evaluate than the other.

In certain cases it is advisable to change to polar coordinates r, θ. Here we divide the field of integration into small areas by means of lines through the origin and circles with centre the origin. The area of a typical small portion is approximately $(r\Delta\theta)(\Delta r)$ and when we take the limit of $\sum$ for this method of division, we get

$$\iint f \, r d\theta dr, \qquad (9.3)$$

where f is expressed in terms of r and θ.

In all problems on the evaluation of double integrals it is advisable to make a sketch of the field of integration.

WORKED EXAMPLES

9.1. *Express as a repeated integral the double integral of $f(x, y)$ over the rectangle bounded by the lines $x = a$, $x = b$, $y = c$, $y = d$. Discuss the case where $f(x, y)$ is of the form $\phi(x)\psi(y)$.*

Using the notation of Fig. 6, we see that $\phi_1(x)$, $\phi_2(x)$, $\psi_1(y)$, $\psi_2(y)$ have the constant values c, d, a, b respectively and the repeated integrals are

$$\int_a^b dx \int_c^d f(x, y) dy, \qquad \int_c^d dy \int_a^b f(x, y) dx.$$

This is known as the case of *constant limits of integration*.

In the special case where $f = \phi(x)\psi(y)$ the double integral is

$$\int_a^b \phi(x) \, dx \int_c^d \psi(y) \, dy,$$

the product of two single integrals. This is known as the case of *variables separable* with constant limits.

9.2. *Evaluate*

$$\int_0^1 dx \int_x^1 \sin\left(\frac{\pi y^2}{2}\right) dy.$$

This is an example where, because of the difficulty of evaluating the inner integral, the order of integration should be changed. First of all the field of integration, which is the triangle bounded by the lines $y = x$, $y = 1$, $x = 0$, should be sketched.

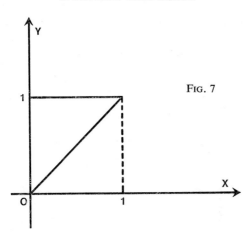

Fig. 7

After change of order the integral is

$$\int_0^1 \sin\left(\frac{\pi y^2}{2}\right) dy \int_0^y dx = \int_0^1 \sin\left(\frac{\pi y^2}{2}\right) y\,dy,$$

which, on letting $u = \pi y^2/2$, becomes

$$\frac{1}{\pi} \int_0^{\pi/2} \sin u\, du = \frac{1}{\pi}.$$

9.3. *Evaluate $\iint xy\,dxdy$, where the field of integration is the area in the first quadrant bounded by the parabolas $y^2 = x$ and $x^2 = y$.*

The integral (Fig. 8, p. 66) is

$$\int_0^1 dx \int_{x^2}^{\sqrt{x}} xy\,dy = \int_0^1 x\left[\frac{y^2}{2}\right]_{x^2}^{\sqrt{x}} dx$$

$$= \int_0^1 \tfrac{1}{2}(x^2 - x^5)\,dx = \tfrac{1}{2}(\tfrac{1}{3} - \tfrac{1}{6}) = \tfrac{1}{12}.$$

9.4. *Find the volume of the solid figure bounded by the cone $z^2 = xy$ and the planes $x = \pm 1, y = \pm 1$.*

From symmetry we see that the required volume is four times the

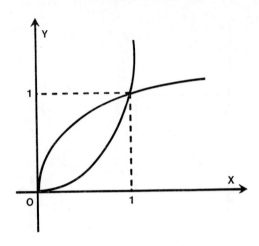

FIG. 8

volume of the solid in the first octant bounded by the planes $z = 0$, $x = 0$, $x = 1$, $y = 0$, $y = 1$ and the cone $z = x^{1/2}y^{1/2}$, i.e. $4 \iint x^{1/2}y^{1/2} \, dx \, dy$ over the square $x = 0$, $x = 1$, $y = 0$, $y = 1$. This is a case of variables separable with constant limits of integration, i.e.

$$4 \int_0^1 y^{1/2} \, dy \int_0^1 x^{1/2} \, dx = 4 \cdot \tfrac{2}{3} \cdot \tfrac{2}{3} = \tfrac{16}{9} \text{ cubic units.}$$

9.5. *Find the volume of a sphere of radius a.*

We calculate the volume of the sphere $x^2 + y^2 + z^2 = a^2$ which lies in the first octant. This volume is the double integral $\iint \sqrt{(a^2 - x^2 - y^2)} \, dx \, dy$ over the first quadrant of the circle $x^2 + y^2 = a^2$ and its value is most easily calculated by changing to polar coordinates. The integral becomes $\iint \sqrt{(a^2 - r^2)} r \, d\theta dr$ and the limits of integration are for θ from 0 to $\tfrac{1}{2}\pi$ and for r from 0 to a, i.e., we have

$$\int_0^{\pi/2} d\theta \int_0^a \sqrt{(a^2 - r^2)} \, r dr = \frac{\pi}{2}\left[-\tfrac{1}{3}(a^2 - r^2)^{3/2} \right]_0^a = \frac{\pi a^3}{6}.$$

The volume of the sphere is eight times this value, i.e., $\tfrac{4}{3}\pi a^3$.

9.6. *Evaluate $\iint (x^2 + y^2)^{-1/2} \tan^{-1}(y/x) dx \, dy$, where the field of*

integration is the area in the first quadrant which lies inside the circle
$x^2 + y^2 = 2x$ *and outside the circle* $x^2 + y^2 = 1$.

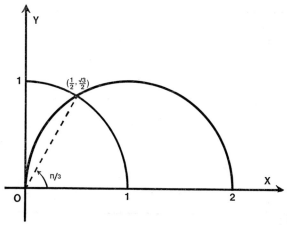

FIG. 9

The circles meet in the point $(\frac{1}{2}, \frac{1}{2}\sqrt{3})$ and they have the polar
equations $r = 2 \cos \theta$, $r = 1$. On changing to polars the integral
becomes

$$\iint r^{-1}\theta r \, d\theta \, dr = \int_0^{\pi/3} \theta d\theta \int_1^{2 \cos \theta} dr$$

$$= \int_0^{\pi/3} \theta(2 \cos \theta - 1) d\theta$$

$$= \left[\theta(2 \sin \theta - \theta) \right]_0^{\pi/3} - \int_0^{\pi/3} (2 \sin \theta - \theta) d\theta$$

$$= \frac{2\pi}{3} \cdot \frac{\sqrt{3}}{2} - \frac{\pi^2}{9} - \left[-2 \cos \theta - \tfrac{1}{2}\theta^2 \right]_0^{\pi/3}$$

$$= \frac{\pi}{\sqrt{3}} - \frac{\pi^2}{9} + 1 + \frac{1}{2} \cdot \frac{\pi^2}{9} - 2 = \frac{\pi}{\sqrt{3}} - \frac{\pi^2}{18} - 1.$$

9.7. *Evaluate*

$$\int_0^1 dx \int_x^{1/x} \frac{y^2 dy}{(x+y)^2 \sqrt{(y^2+1)}}.$$

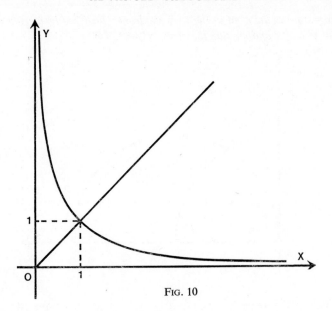

FIG. 10

The integral is most easily evaluated by changing the order of integration. When this is done it becomes clear that we have to divide the area of integration, i.e. the area bounded by the y-axis, the line $y = x$ and the hyperbola $y = 1/x$, into two parts, viz. the triangle bounded by the y-axis and the lines $y = 1$, $y = x$ and the area bounded by the hyperbola, the y-axis and the line $y = 1$. Hence the integral is equal to

$$\int_0^1 \frac{y^2 dy}{\sqrt{(y^2+1)}} \int_0^y \frac{dx}{(x+y)^2} + \int_1^\infty \frac{y^2 dy}{\sqrt{(y^2+1)}} \int_0^{1/y} \frac{dx}{(x+y)^2}$$

$$= \int_0^1 \frac{y^2}{\sqrt{(y^2+1)}} \left[-\frac{1}{x+y} \right]_0^y dy + \int_1^\infty \frac{y^2}{\sqrt{(y^2+1)}} \left[-\frac{1}{x+y} \right]_0^{1/y} dy$$

$$= \tfrac{1}{2} \int_0^1 \frac{y dy}{\sqrt{(y^2+1)}} + \int_1^\infty \frac{y dy}{(1+y^2)^{3/2}}$$

$$= \tfrac{1}{2} \left[\sqrt{(y^2+1)} \right]_0^1 + \left[-\frac{1}{\sqrt{(1+y^2)}} \right]_1^\infty = \sqrt{2} - \tfrac{1}{2}.$$

9.8. *Show that the mean value of the inverse of the distance from the centre of a square of side a of points inside and on the sides of the square is* $(4/a) \log (1+\sqrt{2})$.

The mean value of the function $f(x, y)$ over an area A of the x, y plane is defined to be $\iint_A f(x, y) \, dx \, dy$ divided by the area A.

In our problem we take the square to be that bounded by the lines $x = \pm\frac{1}{2}a$, $y = \pm\frac{1}{2}a$ and the mean value is $\displaystyle\iint \frac{dx \, dy}{\sqrt{(x^2+y^2)}}$ divided by a^2, the area of the square. From symmetry it is clear that this is equal to the double integral over the triangle bounded by $y = 0$, $x = \frac{1}{2}a$ and $y = x$, divided by $\frac{1}{8}a^2$, the area of the triangle. We change to polars and obtain the mean value as

$$\frac{8}{a^2} \int_0^{\pi/4} d\theta \int_0^{(a \sec \theta)/2} dr = \frac{4}{a} \int_0^{\pi/4} \sec \theta \, d\theta$$

$$= \frac{4}{a} \left[\log (\sec \theta + \tan \theta) \right]_0^{\pi/4} = \frac{4}{a} \log (1+\sqrt{2}).$$

9.9 *Show that*

$$\iint (lx+my)^{2n} \, dx \, dy = \frac{\pi(l^2+m^2)^n \, (2n)!}{(n+1)! \, n! \, 2^{2n}}$$

where n is a positive integer and the field of integration is the area within the circle $x^2+y^2 = 1$.

The perpendicular distance p of the point (x, y) from the line $lx+my = 0$ is $|lx+my| / \sqrt{(l^2+m^2)}$. Thus the given integral is

$$(l^2+m^2)^n \iint p^{2n} \, dx \, dy.$$

Now from the symmetry of the circle this is equal to

$$(l^2+m^2)^n \iint y^{2n} \, dx \, dy$$

$$= (l^2+m^2)^n \int_0^{2\pi} \sin^{2n} \theta \, d\theta \int_0^1 r^{2n+1} \, dr$$

$$= 4(l^2+m^2)^n \frac{(2n-1)(2n-3) \ldots 3 . 1}{2n(2n-2) \ldots 4 . 2} \frac{\pi}{2} \frac{1}{2n+2}$$

$$= \frac{\pi(l^2+m^2)^n \, (2n)!}{(n+1)! \, n! \, 2^{2n}}.$$

9.10. *A solid right circular cone has height h and the density at any point of it is proportional to the distance of the point from the axis of the cone. Prove that the centroid of the cone lies on the axis at a distance $\frac{1}{5}$ h from the base.*

From symmetry it is clear that the centroid lies on the axis of the cone. Take the vertex at the origin and let the positive z-axis be the axis of the cone. The density ρ at the point (x, y, z) is given by $\rho = k\sqrt{(x^2+y^2)}$, where k is a constant, and the radius of a section of the cone through this point parallel to the base of radius a is az/h. The mass of a slice of the cone of depth Δz through the point (x, y, z) parallel to the base is

$$\left(\iint k\sqrt{(x^2+y^2)}\,dxdy \right) \Delta z$$

over the above section, which is given by

$$\left(\int_0^{2\pi} d\theta \int_0^R kr^2 dr \right) \Delta z,$$

where $R = az/h$. We sum up all these slices and obtain the mass of the cone as

$$\int_0^h \frac{2\pi k}{3} \left(\frac{az}{h} \right)^3 dz = \frac{\pi a^3 hk}{6}.$$

The z-coordinate of the centroid is

$$\int_0^h \frac{2\pi k}{3} \left(\frac{az}{h} \right)^3 z\,dz$$

divided by the mass of the cone, i.e., $\frac{2}{15}\pi a^3 h^2 k$ divided by $\frac{1}{6}\pi a^3 kh$, i.e. $\frac{4}{5}h$. Hence the centroid lies at a distance $\frac{1}{5}h$ from the base.

ADDITIONAL EXAMPLES

9.11. Evaluate

$$\int_0^{\pi/2} dx \int_x^{\pi/2} \frac{\sin y}{x+y}\,dy.$$

9.12. Evaluate $\iint \dfrac{dx\,dy}{y^2+4ax}$ $(a > 0)$, where the field of integration is the region bounded by the y-axis, the line $y = 1$ and the parabola $y^2 = 4ax$.

9.13. Find the volume enclosed by the cylinder $x^2+y^2-y = 0$, the plane $z = 0$ and the surface $(x^2+y^2)z = y^2$.

9.14. Evaluate $\iint (x+2y^2+y^4)^{-1/2}\,dx\,dy$, where the field of integration is the region bounded by the parabola $y^2 = x$ and the line $x = 1$.

9.15. Evaluate $\iint \dfrac{x\log(x^2+y^2)}{y}\,dx\,dy$, where the field of integration is the area in the first quadrant which lies inside the circle $x^2+y^2 = 2y$ and outside the circle $x^2+y^2 = 1$.

9.16. Evaluate $\iint x^6 e^{xy}\,dx\,dy$, where the field of integration is the area contained by the curves $y = x^2$ and $y = x^5$.

9.17. Show that the cylinders

$$x^2+y^2 = a^2, \quad x^2+(y\cos\alpha-z\sin\alpha)^2 = a^2$$

have their axes inclined at an angle α, and prove that the volume of the solid bounded by the cylinders is $\frac{16}{3}a^3\operatorname{cosec}\alpha$.

9.18. Evaluate

$$\iint \frac{\tan^{-1}(y/x)}{\sqrt{(x^2+y^2)}}\,dx\,dy,$$

where the field of integration is the square bounded by the lines $x = 0, a, y = 0, a$ $(a > 0)$.

9.19. Find the mean value of the square of the distance of points inside and on a circle of radius a from a point distance c from the centre of the circle.

9.20. Evaluate

$$\int_0^1 dx \int_0^{\sqrt{2}} \frac{dy}{(x^2+y^2+1)^2}.$$

Solutions

9.11. $\log 2$.   **9.12.** $(\log 2)/4a$.   **9.13.** $\frac{3}{16}\pi$.   **9.14.** $4(\sqrt{3}-\frac{4}{3})$.

9.15. $\frac{1}{2}(5\log 2 - 3)$.   **9.16.** $\frac{1}{6}(3-e)$.   **9.18.** $\frac{1}{2}\pi a \log(\sqrt{2}+1)$.

9.19. $c^2 + \frac{1}{2}a^2$.   **9.20.** $\frac{1}{4}\pi(\sqrt{2}/4 + \sqrt{2}/3\sqrt{3})$.

10. Triple Integrals

It is easy to extend the process of double integration to that of triple integration. Here the function to be integrated is $f(x, y, z)$ and the 'field' of integration is now a closed volume K in x, y, z space. We divide K into small cuboids with faces parallel to the coordinate planes, a typical cuboid containing the point (x, y, z) having edges of length $\Delta x, \Delta y, \Delta z$. We take the limit of the sum

$$\Sigma = \sum f(x, y, z)\Delta x \Delta y \Delta z,$$

over all the cuboids as their number tends to infinity, where each $\Delta x, \Delta y, \Delta z$ tends to zero. If this limit exists, it is the triple integral of $f(x, y, z)$ throughout K and is denoted by

$$\iiint_K f(x, y\ z)\ dx\ dy\ dz.$$

In the special case where $f \equiv 1$, it measures the volume of K. Exactly as in the case of the double integral this triple integral can be expressed as a repeated integral; for example, as

$$\int_a^b dx \int_{\phi_1(x)}^{\phi_2(x)} dy \int_{\psi_1(x, y)}^{\psi_2(x, y)} f(x, y, z)dz.$$

There are six ways in which we can express the triple integral as a repeated integral, according to the order in which we integrate with respect to the variables x, y, z.

We change to spherical polar coordinates, r, θ, ϕ, as defined in §1, by dividing K into small volumes by spheres $r = $ constant, cones $\theta = $ constant and planes $\phi = $ constant. The volume of such a small region is approximately $(\Delta r)\,(r\Delta\theta)\,(r\sin\theta\Delta\phi)$ and we see that the triple integral becomes

$$\iiint_K f\,r^2\sin\theta\,dr\,d\theta\,d\phi$$

where we express f in terms of r, θ, ϕ.

The n-fold integral

$$\iint\cdots\int f(x_1, x_2, \ldots, x_n)\,dx_1\,dx_2\ldots dx_n$$

throughout a portion of $x_1, x_2, \ldots, x_n$ space is an extension of the above definition.

WORKED EXAMPLES

10.1. *Integrate xyz throughout the volume in the first octant bounded by the coordinate planes and the sphere $x^2+y^2+z^2 = 1$.*

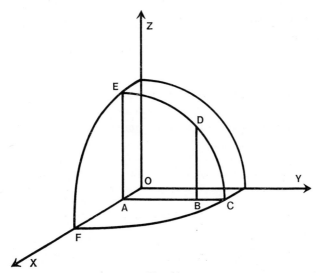

FIG. 11

We shall integrate first with respect to z, then y and finally x. We take a line parallel to the z axis through the point $B(x, y, 0)$ meeting the sphere in $D[x, y, \sqrt{(1-x^2-y^2)}]$. Thus the limits of integration with respect to z are 0 and $\sqrt{(1-x^2-y^2)}$. In the next integral x is kept constant and the limits of integration are y_A and y_C, i.e. 0 and $\sqrt{(1-x^2)}$. In the final integral the limits of integration are x_0 and x_F, i.e. 0 and 1. Thus the integral is

$$\int_0^1 dx \int_0^{\sqrt{(1-x^2)}} dy \int_0^{\sqrt{(1-x^2-y^2)}} xyz\, dz$$

$$= \int_0^1 dx \int_0^{\sqrt{(1-x^2)}} xy \left[\frac{z^2}{2}\right]_{z=0}^{z=\sqrt{(1-x^2-y^2)}} dy$$

$$= \tfrac{1}{2} \int_0^1 dx \int_0^{\sqrt{(1-x^2)}} xy\,(1-x^2-y^2)\, dy$$

$$= \tfrac{1}{2} \int_0^1 x \left[(1-x^2)\frac{y^2}{2}-\frac{y^4}{4}\right]_{y=0}^{y=\sqrt{(1-x^2)}} dx$$

$$= \tfrac{1}{8} \int_0^1 (x-2x^3+x^5)\, dx = \tfrac{1}{48}.$$

An alternative method of evaluating the integral is to use spherical polar coordinates. We then have the integral given by

$$\iiint (r \sin\theta \cos\phi)(r \sin\theta \sin\phi)(r \cos\theta)\, r^2 \sin\theta\, dr\, d\theta d\phi$$

$$= \iiint r^5 \sin^3\theta \cos\theta \cos\phi \sin\phi\, dr\, d\theta\, d\phi.$$

The limits of integration are for r from 0 to 1, for θ from 0 to $\tfrac{1}{2}\pi$ and for ϕ from 0 to $\tfrac{1}{2}\pi$, so that the integral is

$$\int_0^{\pi/2} \cos\phi \sin\phi\, d\phi \int_0^{\pi/2} \sin^3\theta \cos\theta\, d\theta \int_0^1 r^5 dr$$

$$= \frac{1}{2} \cdot \frac{2}{4.2} \cdot \frac{1}{6} = \frac{1}{48}.$$

10.2. *Evaluate $\iiint xy^2z^3 dxdydz$ throughout the volume of the tetrahedron bounded by the planes $z = 0$, $z = x$, $y = 2x$, $y = 2$.*

$$\text{Integral} = \int_0^1 x\,dx \int_{2x}^2 y^2\,dy \int_0^x z^3\,dz$$

$$= \tfrac{1}{4} \int_0^1 x^5\,dx \int_{2x}^2 y^2\,dy$$

$$= \tfrac{2}{3} \int_0^1 x^5(1-x^3)\,dx = \tfrac{1}{27}.$$

10.3. *Evaluate* $\iiint x^2y^2z \, dxdydz$ *throughout the solid figure bounded by the cone* $x^2+y^2 = xz$ *and the plane* $z = 1$.

In spherical polar coordinates the cone has equation $\tan \theta = \cos \phi$ and the plane has equation $r = \sec \theta$. Hence the integral is

$$\int_{-\pi/2}^{\pi/2} \sin^2 \phi \cos^2 \phi \, d\phi \int_0^{\tan^{-1}(\cos \phi)} \sin^5 \theta \cos \theta \, d\theta \int_0^{\sec \theta} r^7\,dr$$

$$= \tfrac{1}{8} \int_{-\pi/2}^{\pi/2} \sin^2 \phi \cos^2 \phi \, d\phi \int_0^{\tan^{-1}(\cos \phi)} \tan^5 \theta \sec^2 \theta \, d\theta$$

$$= \tfrac{1}{48} \int_{-\pi/2}^{\pi/2} \sin^2 \phi \cos^2 \phi \cos^6 \phi \, d\phi = \tfrac{1}{24} \int_0^{\pi/2} \sin^2 \phi \cos^8 \phi \, d\phi$$

$$= \frac{1}{24} \cdot \frac{7 \cdot 5 \cdot 3}{10 \cdot 8 \cdot 6 \cdot 4 \cdot 2} \frac{\pi}{2} = \frac{7\pi}{3 \cdot 2^{12}}.$$

10.4. *Show that*

$$\iiint \frac{dxdydz}{\sqrt{(x^2+y^2+z^2)}} = \frac{a^2}{96} \{6 \log (2+\sqrt{3}) - \pi\},$$

where integration is throughout the tetrahedron bounded by the planes

$$y = 0, \; y = x, \; z = x, \; z = \tfrac{1}{2}a \quad (a > 0).$$

Hence find the mean value of the inverse of the distance from the centre of a cube of side a *of points inside and on the faces of the cube.*

We note that the spherical polar equations of the planes $y = x$, $z = x$ and $z = \tfrac{1}{2}a$ are respectively $\phi = \tfrac{1}{4}\pi$, $\tan \theta = \sec \phi$ and $r = \tfrac{1}{2}a \sec \theta$. Hence the integral is

$$\int_0^{\pi/4} d\phi \int_0^{\tan^{-1}(\sec\phi)} \sin\theta\, d\theta \int_0^{a\frac{1}{2}a\sec\theta} r\,dr$$

$$= \frac{a^2}{8} \int_0^{\pi/4} d\phi \int_0^{\tan^{-1}(\sec\phi)} \sin\theta \sec^2\theta\, d\theta$$

$$= \frac{a^2}{8} \int_0^{\pi/4} \left[\sec\theta\right]_0^{\tan^{-1}(\sec\phi)} d\phi$$

$$= \frac{a^2}{8} \int_0^{\pi/4} \left\{\sqrt{(1+\sec^2\phi)}-1\right\} d\phi$$

$$= \frac{a^2}{8} \int_0^{1/\sqrt{2}} \frac{\sqrt{(2-u^2)}}{1-u^2}\, du - \frac{\pi a^2}{32}, \quad \text{by letting } u = \sin\phi,$$

$$= \frac{a^2}{8} \int_0^{1/\sqrt{2}} \left\{\frac{1}{\sqrt{(2-u^2)}} + \frac{1}{(1-u^2)\sqrt{(2-u^2)}}\right\} du - \frac{\pi a^2}{32}$$

$$= \pi a^2\left(-\tfrac{1}{32}+\tfrac{1}{48}\right) + \frac{a^2}{8} \int_0^{1/\sqrt{2}} \frac{\dfrac{du}{u^3}}{\left(\dfrac{1}{u^2}-1\right)\sqrt{\left(\dfrac{2}{u^2}-1\right)}}$$

$$= -\frac{\pi a^2}{96} + \frac{a^2}{8} \int_{\sqrt{3}}^{\infty} \frac{dv}{v^2-1}, \quad \text{by letting } v^2 = \frac{2}{u^2}-1$$

$$= -\frac{\pi a^2}{96} + \frac{a^2}{16}\left[\log\frac{v-1}{v+1}\right]_{\sqrt{3}}^{\infty} = \frac{a^2}{96}\left\{6\log(2+\sqrt{3})-\pi\right\}.$$

The mean value of $f(x, y, z)$ throughout a volume K is defined to be $\iiint_K f(x, y, z)\, dx\,dy\,dz$ divided by the volume of K.

In our problem we see from symmetry that since the cube consists of 48 tetrahedra, all similarly situated with respect to the origin, the mean value of the inverse of the distance from the centre of a cube of side a of points inside and on the faces is

$$\frac{1}{a^3} \cdot \frac{48a^2}{96}\left\{6\log(2+\sqrt{3})-\pi\right\} = \frac{1}{2a}\left\{6\log(2+\sqrt{3})-\pi\right\}.$$

10.5. *Evaluate*

$$\int_0^a \frac{dx}{(a-x)^{1/2}} \int_0^x \frac{dy}{(x-y)^{1/2}} \int_0^y \frac{e^z}{(x-z)^{1/2}(y-z)^{1/2}}\, dz, \quad (a > 0).$$

Interchange the order in the two innermost integrals and we get

$$\int_0^a \frac{dx}{(a-x)^{1/2}} \int_0^x \frac{e^z dz}{(x-z)^{1/2}} \int_z^x \frac{dy}{(x-y)^{1/2}(y-z)^{1/2}}$$

$$= \int_0^a \frac{dx}{(a-x)^{1/2}} \int_0^x \frac{e^z dz}{(x-z)^{1/2}} \int_0^{\pi/2} 2 d\theta, \quad \text{by letting}$$

$$y = z\cos^2\theta + x\sin^2\theta,$$

$$= \pi \int_0^a \frac{dx}{(a-x)^{1/2}} \int_0^x \frac{e^z dz}{(x-z)^{1/2}}.$$

We now change the order of integration to obtain

$$\pi \int_0^a e^z dz \int_z^a \frac{dx}{(a-x)^{1/2}(x-z)^{1/2}}$$

$$= \pi \int_0^a e^z dz \int_0^{\pi/2} 2 d\theta, \quad \text{by letting } x = z\cos^2\theta + a\sin^2\theta,$$

$$= \pi^2 \int_0^a e^z dz = \pi^2 (e^a - 1).$$

10.6. *Evaluate*

$$\iiiint\limits_K e^{-(x^2 + y^2 + z^2 + u^2)}\, dx\, dy\, dz\, du,$$

where K is the region in four dimensions bounded by
$x^2 + y^2 + z^2 + u^2 = a^2$.

The integral is

$$16 \int_0^a e^{-x^2} dx \int_0^{\sqrt{(a^2 - x^2)}} e^{-y^2} dy \int_0^{\sqrt{(a^2 - x^2 - y^2)}} e^{-z^2} dz$$

$$\int_0^{\sqrt{(a^2 - x^2 - y^2 - z^2)}} e^{-u^2} du.$$

The two innermost integrals are

$$\iint e^{-(z^2+u^2)} \, dz \, du$$

over the first quadrant of the circle $z^2+u^2 = a^2-x^2-y^2$ in the $z\,u$-plane. On changing to polars this is

$$\int_0^{\pi/2} d\theta \int_0^{\sqrt{(a^2-x^2-y^2)}} e^{-r^2} r \, dr = \frac{\pi}{4}(1-e^{x^2+y^2-a^2}).$$

Hence the given four-fold integral is

$$4\pi \iint (e^{-x^2-y^2} - e^{-a^2}) \, dx \, dy$$

over the first quadrant of the circle $x^2+y^2 = a^2$ in the $x\,y$-plane, i.e.,

$$4\pi \int_0^{\pi/2} d\theta \int_0^a (e^{-r^2} - e^{-a^2}) r \, dr = \pi^2(1 - e^{-a^2} - a^2 e^{-a^2}).$$

ADDITIONAL EXAMPLES

10.7. Evaluate

$$\iiint_K xy^2 z^3 \, dx \, dy \, dz,$$

where K is the volume in the first octant bounded by the sphere $x^2+y^2+z^2 = 1$ and the coordinate planes.

10.8. Evaluate

$$\iiint_K xyz \, dx \, dy \, dz,$$

where K is the volume in the first octant bounded by the sphere $x^2+y^2+z^2 = a^2$ and the planes $y = 0$, $z = 0$, $x = y$.

10.9. Prove that

$$\iiint e^{x+2y+3z} \, dx \, dy \, dz = \tfrac{1}{6}(e-1)^3,$$

where integration is throughout the region bounded by the plane $x = y = z = 0$, $x+y+z = 1$.

10.10. Find the mean value of x^2 throughout the positive octant of the solid bounded by the surface

$$x^{2/3}+y^{2/3}+z^{2/3} = a^{2/3} \quad (a > 0).$$

Deduce the value of the moment of inertia about the z-axis of a solid of uniform density occupying this volume.

10.11. Show that

$$\iiint z \, dxdydz = \frac{9\pi a^4}{8},$$

where integration is throughout the volume inside the sphere $x^2+y^2+z^2 = 2az$ $(a>0)$ but outside the sphere $x^2+y^2+z^2 = a^2$.

10.12. Show that the mean squared distance of points inside a sphere of radius a from a tangent plane to the sphere is $\frac{6}{5}a^2$.

10.13. If C is a point distant c $(c > a)$ from the centre of a sphere of radius a, find the mean value of the inverse of the square of the distance from C to points inside the sphere.

10.14. A cone has vertex the origin, axis the z-axis, height h and radius of base a. It is made of material whose density σ obeys the law

$$\sigma = \sigma_0 \left(1+\frac{\rho}{c}\right),$$

where σ_0, c are constants and $\rho^2 = x^2+y^2$. If M is the mass of the cone and C is the moment of inertia about the z-axis, show that $C = \lambda M a^2$, where $5(a+2c)\lambda = 2a+3c$.

SOLUTIONS

10.7. $2/945$.   **10.8.** $a^6/96$.

10.10. $7a^2/143$, $14a^2M/143$, where $M = $ mass.

10.13. $3[2ca-(c^2-a^2) \log \{(c+a)/(c-a)\}]/4a^3c$.

11. Change of Variable in Multiple Integrals

We wish to change the variables in the double integral $\iint_A f(x, y)\,dxdy$ from x, y to u, v, where $x = x(u, v)$, $y = y(u, v)$. It is assumed that these relations can be solved for u, v in terms of x, y and that there is one to one correspondence between points in the x, y plane and points in the uv-plane. The area A in the xy-plane corresponds to an area A' in the uv-plane. It can be shown that the above integral is equal to

$$\iint_{A'} f \left| \frac{\partial(x, y)}{\partial(u, v)} \right| du\,dv,$$

where f is expressed in terms of u, v and $\partial(x, y)/\partial(u, v)$ is the Jacobian defined in §5.

In the case of triple integrals a corresponding result holds:

$$\iiint_K f\,dxdydz = \iiint_{K'} f \left| \frac{\partial(x, y, z)}{\partial(u, v, w)} \right| dudvdw,$$

where the volume K' in u, v, w space corresponds to the volume K in x, y, z space. The method is quite general, i.e.,

$$\iint \cdots \int f\,dx_1 dx_2 \ldots dx_n$$

$$= \iint \cdots \int f \left| \frac{\partial(x_1, x_2, \ldots, x_n)}{\partial(u_1, u_2, \ldots, u_n)} \right| du_1\,du_2 \ldots du_n,$$

where f is a function of the n variables $x_1, x_2, \ldots, x_n$.

It should be noted, as can be easily verified, that the changes to polar and to spherical coordinates are special cases of this general theory.

WORKED EXAMPLES

11.1. *Show that (i) $\iint f(x, y)\,dxdy$ taken over the first quadrant of the ellipse $x^2/a^2 + y^2/b^2 = 1$ is equal to $ab \iint f(a\xi, b\eta)\,d\xi d\eta$ over the first quadrant of the circle $\xi^2 + \eta^2 = 1$; (ii) $\iiint f(x, y, z)\,dxdydz$ throughout the first octant of the ellipsoid $x^2/a^2 + y^2/b^2 + z^2/c^2 = 1$*

is equal to abc $\iiint f(a\xi, b\eta, c\zeta)\, d\xi d\eta d\zeta$ throughout the first octant of the sphere $\xi^2+\eta^2+\zeta^2 = 1$.

(*i*) For the double integral let $x = a\xi$, $y = b\eta$ so that the ellipse becomes the circle $\xi^2+\eta^2 = 1$ while

$$\frac{\partial(x, y)}{\partial(\xi, \eta)} = \begin{vmatrix} \dfrac{\partial x}{\partial \xi} & \dfrac{\partial x}{\partial \eta} \\ \dfrac{\partial y}{\partial \xi} & \dfrac{\partial y}{\partial \eta} \end{vmatrix}$$

$$= \begin{vmatrix} a & 0 \\ 0 & b \end{vmatrix}$$

$$= ab.$$

The result follows at once.

(*ii*) For the triple integral let $x = a\xi$, $y = b\eta$, $z = c\zeta$ and the ellipsoid becomes the sphere $\xi^2+\eta^2+\zeta^2 = 1$ while $|\partial(x, y, z)/\partial(\xi, \eta, \zeta)|$ is equal to abc. The result follows at once.

11.2. *The cross section of a cylinder whose generators are parallel to the z-axis is the area in the first quadrant of the xy-plane that is bounded by the hyperbolas $x^2-y^2 = a^2$, $x^2-y^2 = 2a^2$, $xy = a^2$, $xy = 2a^2$ ($a > 0$). Show that the volume of this cylinder intercepted between the plane $z = 0$ and the paraboloid $az = x^2+y^2$ is $\frac{1}{2}a^3$.*

Let $u = x^2-y^2$, $v = xy$, so that

$$\frac{\partial(u, v)}{\partial(x, y)} = \begin{vmatrix} 2x & -2y \\ y & x \end{vmatrix} = 2(x^2+y^2)$$

and hence $\dfrac{\partial(x, y)}{\partial(u, v)} = \dfrac{1}{2(x^2+y^2)}$. The required volume is

$$\iint \left(\frac{x^2+y^2}{a}\right) dxdy$$ over the area in the first quadrant bounded by the four hyperbolas. Under the above transformation the hyperbolas become the straight lines $u = a^2$, $u = 2a^2$, $v = a^2$, $v = 2a^2$ and so the field of integration in the uv-plane is the rectangle bounded by these lines. The volume is therefore $\displaystyle\iint \left(\frac{x^2+y^2}{a}\right) \frac{1}{2(x^2+y^2)}\, dudv$ over this rectangle, i.e.

$$\frac{1}{2a} \iint dudv = \frac{1}{2a}\text{ (area of the rectangle)} = \frac{1}{2a}a^4 = \frac{1}{2}a^3$$

11.3. *Evaluate*

$$\iint \frac{dxdy}{\sqrt{(x^2+xy+y^2)}},$$

where the field of integration is the triangle bounded by the coordinate axes and the line $2x+y = 2$.

Let $u = 2x+y$, $v = y\sqrt{3}$, so that $u^2+v^2 = 4(x^2+y^2+xy)$ and $\partial(u, v)/\partial(x, y) = 2\sqrt{3}$. The boundaries of the field are the lines $2x+y = 2$, $y = 0$, $x = 0$. Under the transformation these become $u = 2$, $v = 0$, $v = u\sqrt{3}$, so that the field in the uv-plane is the triangle bounded by these lines.

The integral is

$$\frac{1}{2\sqrt{3}} \iint \frac{2\,dudv}{\sqrt{(u^2+v^2)}}$$

over this triangle. We now change to polar coordinates to obtain the integral as

$$\frac{1}{\sqrt{3}} \int_0^{\pi/3} d\theta \int_0^{2\sec\theta} dr = \frac{2}{\sqrt{3}} \int_0^{\pi/3} \sec\theta\,d\theta$$

$$= \left[\frac{2}{\sqrt{3}} \log(\sec\theta+\tan\theta) \right]_0^{\pi/3} = \frac{2}{\sqrt{3}} \log(2+\sqrt{3}).$$

11.4. *Evaluate*

$$\iint e^{-\sqrt{(ax^2+2hxy+by^2)}}\,dxdy,$$

where the field of integration is the interior of the ellipse $ax^2+2hxy+by^2 = 1$ $(a > 0, ab-h^2 > 0)$.

We rotate the axes through an angle θ so that the new X, Y-axes are along the axes of the ellipse; we have

$$x = X\cos\theta - Y\sin\theta, \quad y = X\sin\theta + Y\cos\theta$$

and $\partial(x, y)/\partial(X, Y) = \cos^2\theta+\sin^2\theta = 1$. Under this transformation $ax^2+2hxy+by^2 = \lambda_1 X^2+\lambda_2 Y^2$, where $\lambda_1\lambda_2 = ab-h^2$. The integral is now

$$\iint e^{-\sqrt{(\lambda_1 X^2+\lambda_2 Y^2)}}\,dXdY$$

over the ellipse $\lambda_1 X^2 + \lambda_2 Y^2 = 1$. We next let $\xi = \sqrt{\lambda_1} X$, $\eta = \sqrt{\lambda_2} Y$ (we know that $\lambda_1, \lambda_2 > 0$) and the integral becomes $\iint e^{-\sqrt{(\xi^2 + \eta^2)}} (\lambda_1 \lambda_2)^{-1/2} \, d\xi d\eta$ over the circle $\xi^2 + \eta^2 = 1$. The final change of variable is to polar coordinates and we get

$$(ab - h^2)^{-1/2} \iint e^{-r} r \, d\theta \, dr$$

over the circle, i.e.,

$$2\pi (ab - h^2)^{-1/2} \int_0^1 e^{-r} r \, dr = \frac{2\pi}{\sqrt{(ab - h^2)}} \left(1 - \frac{2}{e} \right).$$

11.5. *Find the mean value of the square of the distance from the origin of points in the volume in the first octant bounded by the planes* $x = 0$, $x - y = 0$, *the cones* $x^2 + y^2 - z^2 = 0$, $x^2 + y^2 - 2z^2 = 0$ *and the planes* $z = 1$, $z = 4$.

We make the transformation

$$u = \frac{x}{y}, \quad v = \frac{z^2}{x^2 + y^2 + z^2}, \quad w = z$$

and a simple calculation gives

$$\frac{\partial(u, v, w)}{\partial(x, y, z)} = \frac{2v^2(1 + u^2)}{w^2}.$$

The boundaries of the volume $x = 0$, $x - y = 0$, $x^2 + y^2 - z^2 = 0$, $x^2 + y^2 - 2z^2 = 0$, $z = 1$, $z = 4$ become respectively $u = 0$, $u = 1$, $v = \frac{1}{2}$, $v = \frac{1}{3}$, $w = 1$, $w = 4$; also $x^2 + y^2 + z^2 = w^2/v$. Hence $\iiint (x^2 + y^2 + z^2) \, dx dy dz$ throughout the volume is

$$\iiint \frac{w^2}{v} \cdot \frac{w^2}{2v^2(1 + u^2)} \, du \, dv \, dw,$$

$$= \frac{1}{2} \int_0^1 \frac{du}{1 + u^2} \int_{1/3}^{1/2} \frac{dv}{v^3} \int_1^4 w^4 dw,$$

$$= \frac{1}{2} \cdot \frac{\pi}{4} \cdot \frac{1}{2} (9 - 4) \frac{1}{5} (4^5 - 1) = \frac{1023\pi}{16}.$$

Similarly, the volume is

$$\iiint \frac{w^2}{2v^2(1+u^2)} \, du \, dv \, dw,$$

$$= \frac{1}{2} \int_0^1 \frac{du}{1+u^2} \int_{1/3}^{1/2} \frac{dv}{v^2} \int_1^4 w^2 dw,$$

$$= \frac{1}{2} \cdot \frac{\pi}{4} (3-2) \frac{64-1}{3} = \frac{21\pi}{8}.$$

Hence the required mean value is $(1023\pi/16)(8/21\pi) = 341/14$.

11.6. *Prove that the value of the double integral*

$$\iint e^{-(x^2+y^2+2xy\cos\alpha)} \, dx \, dy \quad (0 < \alpha < \pi)$$

taken over the infinite quadrant $x \geqslant 0$, $y \geqslant 0$ *is* $\alpha/2 \sin \alpha$.

Let $u = x + y \cos \alpha$, $v = y \sin \alpha$, so that $\partial(u, v)/\partial(x, y) = \sin \alpha$. The boundaries $y = 0$, $x = 0$ become $v = 0$, $v = u \tan \alpha$. Hence the field of integration in the uv-plane is the infinite sector bounded by the lines $v = 0$, $v = u \tan \alpha$. Now $x^2 + y^2 + 2xy \cos \alpha = u^2 + v^2$, and on changing to polar coordinates the integral becomes

$$\frac{1}{\sin \alpha} \int_0^\alpha d\theta \int_0^\infty e^{-r^2} r \, dr = \frac{\alpha}{2 \sin \alpha} \left[-e^{-r^2} \right]_0^\infty = \frac{\alpha}{2 \sin \alpha}.$$

ADDITIONAL EXAMPLES

11.7. A is the finite area of the xy-plane bounded by the hyperbolas $xy = 1$, $xy = 2$ and the parabolas $y^2 = x$, $y^2 = 2x$. Show that the volume of the solid bounded by the plane $z = 0$, the surface $z = x^3$ and the cylinder with generators parallel to the z-axis through the boundary of A is $\frac{7}{18}$.

11.8. Show that the area bounded by the circles $x^2 + y^2 = 2x$, $x^2 + y^2 = 4x$ and the lines $y = x$, $y = 0$ is $\frac{3}{4}(\pi + 2)$.

11.9. Show, by using the substitution $u = y^2/x$, $v = x^2/y$, that $\iint xy \, dxdy = \frac{5}{72}$, where the field of integration is the area bounded by the loop of the curve $x^3 + y^3 = xy$ and the arcs of the parabolas $y^2 = x$, $x^2 = y$ joining the origin to the point $(1, 1)$.

11.10. Show that

$$\iiint (x+y+z)^{2n}\, dxdydz = \frac{4\pi abc\,(a^2+b^2+c^2)^n}{(2n+1)(2n+3)}$$

where integration is throughout the volume of the ellipsoid $x^2/a^2+y^2/b^2+z^2/c^2 = 1$ and n is a positive integer.

11.11. Show that the mean value of the function

$$x^3 e^{x^2/a^2+y^2/b^2+z^2/c^2}$$

taken throughout the part of the volume of the ellipsoid $x^2/a^2+y^2/b^2+z^2/c^2 = 1$ which lies in the first octant is $\frac{3}{8}(e-2)a^3$.

11.12. A cylinder has generators parallel to the z-axis and its section by the plane $z = 0$ is the finite area bounded by the arcs $y = x^2$, $2y = x^2$, $x^3y = 1$, $x^3y = 6$. Show that the volume of this cylinder between the plane $z = 0$ and the surface $z = x^2$ is $\log 2$.

Chapter IV

LINE AND SURFACE INTEGRALS

12. Line Integrals Along a Plane Curve

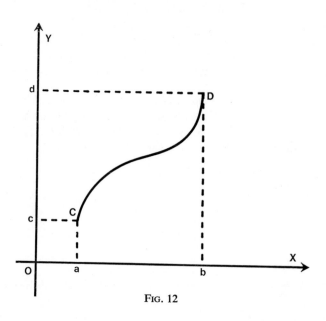

Fig. 12

Let $F(x, y)$ be a function of x, y and let $f(x)$ be a single-valued function of x, represented in Fig. 12 by the curve CD. Then the integral $\int_a^b F[x, f(x)]dx$ is written $\int_{CD} F(x, y)dx$ and is called a curvilinear or line integral along the curve $y = f(x)$ from C to D. Suppose further that the equation of CD can be written in the form $x = \psi(y)$, where $\psi(y)$ is a single-valued function of y; then $\int_c^d F[\psi(y), y] \, dy$ is the line integral $\int_{CD} F(x, y) \, dy$ along the curve from C to D.

86

It is easily verified that

$$\int_{DC} F(x, y)dx = -\int_{CD} F(x, y)dx, \quad \int_{DC} F(x, y)dy = -\int_{CD} F(x, y)dy.$$

$$(12.1)$$

Let the arc CD be divided in any way into small portions, a typical portion having length Δs, and let (x, y) be any point in this small portion of the arc. We form the sum $\sum = \sum F(x, y)\Delta s$ over all these portions and if $\sum$ tends to a limit as each Δs tends to zero, this limit is called the line integral $\int_{CD} F(x, y)ds$. From this definition it is clear that this integral, as distinct from those in (12.1), is independent of the direction along the arc in which it is taken. Since $ds/dx = \sqrt{(1+(dy/dx)^2)}$,

$$\int_{CD} F(x, y)ds = \int_a^b F[x, f(x)]\sqrt{\{1+(f'(x)^2)\}}\, dx,$$

where $b > a$. In this integral integration along the x-axis must be in the positive direction.

In the case of a curve which is the boundary of an area in the plane we distinguish between the directions along the curve by means of the terms positive and negative. We define the positive direction to be such that a person moving round the curve in the

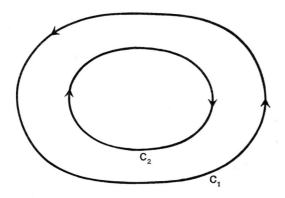

FIG. 13

positive direction has the area enclosed by the curve on his left-hand side.

In Fig. 13 the positive direction round the area between the curves C_1 and C_2 is shown by the arrow heads.

Gauss's (or Green's) theorem states that, if $P(x, y)$, $Q(x, y)$ are functions defined inside and on the boundary C of a closed area K, then

$$\iint_K \left(\frac{\partial Q}{\partial x} - \frac{\partial P}{\partial y}\right) dxdy = \int_C (Pdx + Qdy).$$

WORKED EXAMPLES

12.1. *Evaluate*

$$\int (2xy + y^2)\,dx + (x^2 + xy)\,dy$$

taken positively round the boundary of the region cut off from the first quadrant by the curve $y = 1 - x^3$.

We shall evaluate the integral directly and check the result by means of Gauss's theorem:

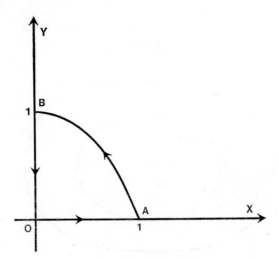

FIG. 14

Along OA $\int (2xy+y^2)dx = 0$, since $y = 0$; $\int (x^2+xy)dy = 0$, since the value of y is constant. Similarly, the integrals along BO are zero. Along AB the integral is

$$\int_1^0 2x(1-x^3)+(1-x^3)^2dx+\int_0^1 (1-y)^{2/3}+y(1-y)^{1/3}dy$$

$$= -\tfrac{87}{70}+\tfrac{129}{140} = -\tfrac{45}{140} = -\tfrac{9}{28}.$$

Using Gauss's Theorem, we take $P = 2xy+y^2$, $Q = x^2+xy$, and so the line integral is equal to the double integral,

$$\iint (2x+y-2x-2y)\,dxdy = \iint -y\,dxdy$$

over the given area OAB. This is

$$\int_0^1 dx \int_0^{1-x^3} (-y)\,dy = -\tfrac{1}{2}\int_0^1 (1-x^3)^2\,dx = -\tfrac{9}{28}.$$

12.2. *If C is the circle $x^2+y^2 = 2x$, evaluate $\displaystyle\int_C x\,ds$.*

From symmetry it is clear that the integral is twice that over the part of the circle in the first quadrant. Here $y = (2x-x^2)^{1/2}$, $y' = (1-x)(2x-x^2)^{-1/2}$, $\sqrt{(1+y'^2)} = (2x-x^2)^{-1/2}$. Hence the required line integral is

$$2\int_0^2 \frac{xdx}{\sqrt{(2x-x^2)}} = 8\int_0^{\pi/2} \sin^2 \theta \, d\theta, \text{ where } x = 2\sin^2 \theta$$

$$= 8\frac{\pi}{4} = 2\pi.$$

12.3. *Show that*

$$\iint_K \nabla^2 V \, dxdy = \int_C \frac{\partial V}{\partial n}\,ds,$$

where C is the boundary of K and $\partial V/\partial n$ is the derivative of V in the direction of the outward normal to C.

As in **2.2,** $\nabla^2 V = \partial^2 V/\partial x^2 + \partial^2 V/\partial y^2.$

D

In Gauss's theorem, let $Q = \partial V/\partial x$, $P = -\partial V/\partial y$, and we have

$$\iint\limits_{K} \nabla^2 V \, dx \, dy = \int\limits_{C} \left(\frac{\partial V}{\partial x} \, dy - \frac{\partial V}{\partial y} \, dx \right), \text{ in the positive direction}$$

$$= \int\limits_{C} \left(\frac{\partial V}{\partial x} \cdot \frac{dy}{ds} - \frac{\partial V}{\partial y} \cdot \frac{dx}{ds} \right) ds,$$

where dy/ds, dx/ds are respectively $\sin \theta$, $\cos \theta$, θ being the angle which the positive direction of the tangent makes with the positive direction of the x-axis. If ϕ is the angle which the outward drawn normal makes with the x-axis, $\phi = \theta - \tfrac{1}{2}\pi$, and so

$$\iint\limits_{K} \nabla^2 V \, dx \, dy = \int\limits_{C} \left(\frac{\partial V}{\partial x} \cos \phi + \frac{\partial V}{\partial y} \sin \phi \right) ds$$

$$= \int\limits_{C} \frac{\partial V}{\partial n} \, ds.$$

(See (4.5).)

12.4. *If $Pdx + Qdy$ is an exact differential, show that* $\displaystyle\int\limits_{CD} (Pdx + Qdy)$ *has the same value for all curves $y = f(x)$ joining C and D, where $f(x)$ is a single-valued function of x.*

For the definition of an exact differential see §5. Let $u(x, y)$ be the function of which $Pdx + Qdy$ is the exact differential. Then

$$\int\limits_{CD} (Pdx + Qdy) = \left[u \right]_C^D = u(x_2, y_2) - u(x_1, y_1),$$

where C, D are $(x_1, y_1), (x_2, y_2)$, for any curve $y = f(x)$ joining C and D.

12.5. *Show that*

(i) $$\int \frac{x \, dy - y \, dx}{x^2 + y^2} = 0,$$

where the line integral is taken around the closed contour consisting

*of the upper halves of the circle $x^2 + y^2 = 9$ and the ellipse $x^2 + 4y^2 = 4$
together with the segments of the x-axis, from $x = -3$ to $x = -2$,
and from $x = 2$ to $x = 3$;*

(ii) $$\int \frac{x\,dy - y\,dx}{x^2 + y^2} = 2\pi,$$

*where integration is round the ellipse $x^2 + 4y^2 = 4$ in the positive
direction.*

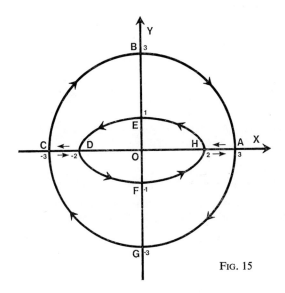

Fig. 15

Let $P = -y/(x^2 + y^2)$, $Q = x/(x^2 + y^2)$. Then

$$\frac{\partial P}{\partial y} = \frac{y^2 - x^2}{(x^2 + y^2)^2} = \frac{\partial Q}{\partial x},$$

and we can apply Gauss's theorem in (i) to show that round the
given contour

$$\int \frac{x\,dy - y\,dx}{x^2 + y^2} = \iint 0 \, dx \, dy = 0.$$

(ii) We cannot apply Gauss's theorem since P and Q are not
defined at the origin. However, using the notation of Fig. 15, we have

D 2

$$\int\limits_{HEDCBAH} \frac{xdy-ydx}{x^2+y^2} + \int\limits_{HAGCDFH} \frac{xdy-ydx}{x^2+y^2} = 0.$$

Since the integrals along the straight segments cancel out, we have the integral round the ellipse in the positive direction equal to the integral round the circle in the positive direction. Since $x^2+y^2 = 9$ on the circle, this latter integral is equal to $\frac{1}{9}\int (xdy-ydx)$ round the circle in the positive direction. Take $P = -y$, $Q = x$ in Gauss's theorem and we get $\frac{2}{9}\iint dxdy$ over the interior of the circle, i.e. $\frac{2}{9}$ (area of the circle) $= \frac{2}{9}(9\pi) = 2\pi$.

12.6. *Prove that*

$$\iint\limits_{K} (u\nabla^2 v - v\nabla^2 u)\,dxdy = \int\limits_{C}\left(u\frac{\partial v}{\partial n} - v\frac{\partial u}{\partial n}\right)ds,$$

where C is the boundary of K and $\delta/\delta n$ denotes differentiation in the direction of the outward normal to C.

Note that $\dfrac{\partial}{\partial x}\left(u\dfrac{\partial v}{\partial x} - v\dfrac{\partial u}{\partial x}\right) = u\dfrac{\partial^2 v}{\partial x^2} - v\dfrac{\partial^2 u}{\partial x^2}$ and similarly for differentiation with respect to y. In Gauss's theorem let $Q = u\partial v/\partial x - v\partial u/\partial x$ and $P = -(u\partial v/\partial y - v\partial u/\partial y)$ and we get

$$\iint\limits_{K} (u\nabla^2 v - v\nabla^2 u)\,dxdy = \int\limits_{C} u\left(\frac{\partial v}{\partial x}\,dy - \frac{\partial v}{\partial y}\,dx\right) - v\left(\frac{\partial u}{\partial x}\,dy - \frac{\partial u}{\partial y}\,dx\right)$$

$$= \int\limits_{C}\left(u\frac{\partial v}{\partial n} - v\frac{\partial u}{\partial n}\right)ds,$$

exactly as in **12.3.**

12.7. *Show that the mean value of* $\log p$, *where p is the length of the perpendicular from a point on a circle of radius 1 to a given tangent to the circle, taken over all points on the circle is* $-\log 2$.

The mean value of a function $f(x, y)$ on an arc of a curve is defined as $\int f(x, y)\,ds$ along the arc, divided by the length of the arc.

In our problem we take the circle to have equation $x^2+y^2-2x = 0$ and the given tangent to be the y-axis so that $p = x$. From symmetry we need only consider the semicircle above the x-axis. The required mean value is

$$\frac{1}{\pi}\int \log x \, ds$$

over this semicircle. Now

$$\left(\frac{ds}{dx}\right)^2 = 1+\left(\frac{dy}{dx}\right)^2 = 1+\frac{(1-x)^2}{2x-x^2} = \frac{1}{2x-x^2}.$$

Hence the mean value is

$$\frac{1}{\pi}\int_0^2 \frac{\log x \, dx}{\sqrt{(2x-x^2)}} = \frac{1}{\pi}\int_0^{\pi/2} \frac{\log (2\sin^2 \theta)}{2\sin\theta\cos\theta} 4\sin\theta\cos\theta \, d\theta,$$

$$\text{where } x = 2\sin^2 \theta,$$

$$= \frac{2}{\pi}\left\{\frac{\pi}{2}\log 2 + 2\int_0^{\pi/2} \log \sin\theta \, d\theta\right\}$$

$$= \frac{2}{\pi}\left\{\frac{\pi}{2}\log 2 - \pi\log 2\right\} = -\log 2.$$

ADDITIONAL EXAMPLES

12.8. Find $\int (x^2+y^2)dx$ and $\int (x^2+y^2)dy$ taken from $(0,0)$ to $(1,1)$ along the parabola $y = x^2$.

12.9. Find $\int y \, ds$ over the upper half of the circle $x^2+y^2 = a^2$.

12.10. Show that

$$\int \frac{xdy-ydx}{x^2-xy+y^2} = \frac{4\pi}{\sqrt{3}},$$

where integration is round the ellipse $x^2+xy+y^2 = 1$ in the positive direction.

12.11. If $\psi = x^4+y^4$ and C is the circle $x^2+y^2 = 1$, evaluate $\int_C \frac{\partial\psi}{\partial n} ds$, (i) directly; (ii) using Gauss's theorem.

12.12. Show that the mean value of the inverse of the distance from the centre of a square of side a of points on the sides of the square is $2a^{-1} \log (1+\sqrt{2})$.

12.13. Find the mean value of the square of the distance of points on the circumference of a circle of radius a from a point distant c $(c > a)$ from the centre of the circle.

12.14. If C is the boundary of a closed area K, if z satisfies the equation $\partial^2 z/\partial x^2 + \partial^2 z/\partial y^2 = 0$ inside and on C, and if $w = 0$ on C, show that

$$\iint\limits_K \left(\frac{\partial z}{\partial x} \frac{\partial w}{\partial x} + \frac{\partial z}{\partial y} \frac{\partial w}{\partial y} \right) dxdy = 0.$$

12.15. The closed curve C consists of the arc of the parabola $y^2 = 4ax$ $(a > 0)$ between the points $(a, 2a)$ and $(a, -2a)$ and the straight line joining $(a, -2a)$ and $(a, 2a)$.

Show that

$$\int\limits_C (x^2 y \, dx + xy^2 dy) = \tfrac{104}{105} a^4,$$

where C is taken in a counterclockwise direction.

SOLUTIONS

12.8. 8/15, 5/6.   **12.9.** $2a^2$.   **12.11.** 6π.

12.13. $c^2 + a^2$.

13. Area of a Surface

Let A be the projection on the plane $z = 0$ of the portion S of the surface with equation $z = z(x, y)$. Then it can be shown that the area of S is given by

$$\iint\limits_A \sqrt{(1 + p^2 + q^2)} \, dxdy, \qquad (13.1)$$

where $p = \partial z/\partial x, q = \partial z/\partial y$.

If the surface is defined by parametric equations

$$x = x(u, v), \quad y = y(u, v), \quad z = z(u, v),$$

the area of a portion of the surface is given by

$$\iint \sqrt{\left\{\left[\frac{\partial(y, z)}{\partial(u, v)}\right]^2 + \left[\frac{\partial(z, x)}{\partial(u, v)}\right]^2 + \left[\frac{\partial(x, y)}{\partial(u, v)}\right]^2\right\}} \, du \, dv \quad (13.2)$$

over the appropriate area in the uv-plane.

WORKED EXAMPLES

13.1. *Find the area of the surface $z = xy$ in the first octant cut off by the cylinder $x^2 + y^2 = 1$.*

Here $p = y$, $q = x$, $\sqrt{(1 + p^2 + q^2)} = \sqrt{(1 + y^2 + x^2)}$ and the required surface area is $\iint \sqrt{(1 + x^2 + y^2)} \, dxdy$ over the first quadrant of the circle $x^2 + y^2 = 1$. On changing to polar coordinates this is

$$\iint \sqrt{(1 + r^2)} \, r \, dr \, d\theta = \int_0^{\pi/2} d\theta \int_0^1 \sqrt{(1 + r^2)} \, r \, dr = \frac{\pi}{2}\left[\tfrac{1}{3}(1 + r^2)^{3/2}\right]_0^1$$

$$= \tfrac{1}{6}\pi(2\sqrt{2} - 1).$$

13.2. *Find the area of the part of the cone $z^2 = 2xy$ which lies inside the sphere $x^2 + y^2 + z^2 = 1$.*

The cone meets the sphere where $x^2 + y^2 + 2xy = 1$, i.e., where $x + y = \pm 1$. Hence, from symmetry, the required surface is

$$4 \iint \sqrt{(1 + p^2 + q^2)} \, dxdy$$

over the triangle bounded by $y = 0$, $x = 0$, $x + y = 1$. Now

$$p = \frac{y}{z}, \quad q = \frac{x}{z},$$

$$\sqrt{(1 + p^2 + q^2)} = \sqrt{\left(1 + \frac{x^2 + y^2}{z^2}\right)} = \sqrt{\frac{(2xy + x^2 + y^2)}{2xy}}$$

and hence the surface area is

$$\frac{4}{\sqrt{2}} \int_0^1 dx \int_0^{1-x} \frac{(x+y)}{\sqrt{(xy)}}\, dy$$

$$= \frac{4}{\sqrt{2}} \int_0^1 \{2x^{1/2}(1-x)^{1/2} + \tfrac{2}{3}x^{-1/2}(1-x)^{3/2}\}\, dx$$

$$= \frac{8}{\sqrt{2}} \int_0^{\pi/2} (2\sin^2\theta\cos^2\theta + \tfrac{2}{3}\cos^4\theta)\, d\theta,$$

$$\text{where } x = \sin^2\theta,$$

$$= \frac{8}{\sqrt{2}} \left(\frac{\pi}{8} + \frac{\pi}{8}\right) = \sqrt{2\pi}.$$

13.3. *Show that the direction cosines of the normal at the point* (x, y, z) *on the surface*

$$\sin z = \sinh x \sinh y, \quad -\tfrac{1}{2}\pi \leqslant z \leqslant \tfrac{1}{2}\pi,$$

are $\tanh y$, $\tanh x$, $-\cos z/(\cosh x \cosh y)$. *Find the area of the surface between the planes* $x = a$, $x = b$, *where* $a > b > 0$.

Let $f = \sinh x \sinh y - \sin z = 0$; then the direction cosines of the normal at (x, y, z) are proportional to f_x, f_y, f_z (see §4). Now

$$f_x = \cosh x \sinh y, \quad f_y = \sinh x \cosh y, \quad f_z = -\cos z$$

and the actual direction cosines are these quantities divided by
$$\sqrt{(\cosh^2 x \sinh^2 y + \sinh^2 x \cosh^2 y + \cos^2 z)}$$
$$= \sqrt{(\cosh^2 x \sinh^2 y + \sinh^2 x \cosh^2 y + 1 - \sinh^2 x \sinh^2 y)}$$
$$= \cosh x \cosh y,$$
that is, the direction cosines are $\tanh y$, $\tanh x$, $-\cos z/(\cosh x \cosh y)$. To evaluate the surface area we project on to the plane $x = 0$, where the projected area is bounded by the curves $\sinh y = \sin z/\sinh b$ and $\sinh y = \sin z/\sinh a$ and the lines $z = -\tfrac{1}{2}\pi$ and $z = \tfrac{1}{2}\pi$. Thus the area is

$$\int_{-\pi/2}^{\pi/2} dz \int_{\sinh^{-1}(\sin z/\sinh a)}^{\sinh^{-1}(\sin z/\sinh b)} dy/|\tanh y|$$

$$= 2\int_0^{\pi/2} \left(\log\frac{\sin z}{\sinh b} - \log\frac{\sin z}{\sinh a}\right) dz = \pi \log\frac{\sinh a}{\sinh b}.$$

13.4. *The circular cylinder with axis the line* $z = x \tan \alpha$, $y = 0$ *($0 < \alpha < \frac{1}{2}\pi$) and radius of cross section r has equation*

$$y^2 + (x \sin \alpha - z \cos \alpha)^2 = r^2.$$

Show that the area of the surface of this cylinder cut off by the cylinder with generators parallel to the z-axis through the intersection of the circular cylinder with the plane $z = 0$ is $16r^2 \operatorname{cosec} 2\alpha$.

The upper half of the circular cylinder has equation

$$z \cos \alpha = x \sin \alpha + \sqrt{(r^2 - y^2)},$$

so that $p = \tan \alpha$, $q = -y/\cos \alpha \sqrt{(r^2 - y^2)}$. Hence

$$\sqrt{(1 + p^2 + q^2)} = \sqrt{\left(1 + \tan^2 \alpha + \frac{y^2 \sec^2 \alpha}{r^2 - y^2}\right)} = \frac{r \sec \alpha}{\sqrt{(r^2 - y^2)}}.$$

Now the cylinder meets the plane $z = 0$ in the ellipse $x^2 \sin^2 \alpha + y^2 = r^2$ and so the area of the surface cut off this top half of the circular cylinder by the cylinder through this ellipse with generators parallel to the z-axis is

$$\iint \frac{r \sec \alpha}{\sqrt{(r^2 - y^2)}} \, dx dy$$

over this ellipse, i.e.,

$$4 \int_0^r r \sec \alpha \, \frac{1}{\sqrt{(r^2 - y^2)}} \, dy \int_0^{\operatorname{cosec} \alpha \sqrt{(r^2 - y^2)}} dx$$

$$= \frac{4r^2}{\sin \alpha \cos \alpha} = 8r^2 \operatorname{cosec} 2\alpha.$$

From symmetry the area cut off the lower half of the circular cylinder has the same value and thus the total area cut off the circular cylinder is $16r^2 \operatorname{cosec} 2\alpha$.

13.5. *The cylinder with equation $r = \sin 2\theta$ ($0 \leqslant \theta \leqslant \frac{1}{2}\pi$) in cylindrical coordinates r, θ, z cuts the paraboloid $z = r^2$ in the curve C. A cone is formed by joining the points of C to the origin. Show that this cone has parametric equations*

$$x = u \cos v, \quad y = u \sin v, \quad z = u \sin 2v.$$

Hence find the area of the part of the cone within the cylinder $r^2 = \sin 2\theta$.

At the intersection of the cylinder and the paraboloid $z/r = \sin 2\theta$. Since this equation is homogeneous in z and r it represents the required cone. We now let $r = u$, $\theta = v$ and get $x = u \cos v$, $y = u \sin v$, $z = u \sin 2v$ as the parametric equations of the cone. A simple calculation shows that

$$\frac{\partial(y, z)}{\partial(u, v)} = u\,(2 \sin v \cos 2v - \sin 2v \cos v)$$

$$\frac{\partial(z, x)}{\partial(u, v)} = -u\,(2 \cos v \cos 2v + \sin 2v \sin v), \text{ and}$$

$$\frac{\partial(x, y)}{\partial(u, v)} = u.$$

Hence

$$\sqrt{\left[\left\{\frac{\partial(y, z)}{\partial(u, v)}\right\}^2 + \left\{\frac{\partial(z, x)}{\partial(u, v)}\right\}^2 + \left\{\frac{(\partial x, y)}{\partial(u, v)}\right\}^2\right]} = u\sqrt{(2 + 3 \cos^2 2v)},$$

so that the surface area of the cone within the cylinder $r^2 = \sin 2\theta$ is

$$\int_0^{\pi/2} dv \int_0^{\sqrt{(\sin 2v)}} \sqrt{(2 + 3 \cos^2 2v)}\,u\,du$$

$$= \tfrac{1}{2} \int_0^{\pi/2} \sqrt{(2 + 3 \cos^2 2v)} \sin 2v\, dv$$

$$= \tfrac{1}{4} \int_{-1}^{1} \sqrt{(2 + 3z^2)}\,dz, \text{ by taking } z = \cos 2v,$$

$$= \tfrac{\sqrt{3}}{2} \left[\tfrac{1}{2}z \sqrt{(z^2 + \tfrac{2}{3})} + \tfrac{1}{3} \log \{z + \sqrt{(z^2 + \tfrac{2}{3})}\}\right]_0^1$$

$$= \tfrac{1}{4}\sqrt{5} + \tfrac{1}{6}\sqrt{3} \log (\sqrt{\tfrac{3}{2}} + \sqrt{\tfrac{5}{2}})$$

$$= \tfrac{1}{4}\sqrt{5} + \tfrac{1}{6}\sqrt{3} \log \tfrac{1}{2}(\sqrt{6} + \sqrt{10}).$$

ADDITIONAL EXAMPLES

13.6. Find the area of the part of the surface of the cylinder $x^2 + z^2 = 2x$ which is bounded by the planes $y = 0$ and $x + y = 2$.

13.7. Find the area of the surface of the cone $x^2 + y^2 = z^2 \tan^2 \alpha$ $(0 < \alpha < \frac{1}{2}\pi)$ which lies inside the sphere $x^2 + y^2 + z^2 = 2ay$ $(a > 0)$.

13.8. Find the area intercepted on the surface of the sphere $x^2 + y^2 + z^2 = a^2$ by the cylinder whose generators are parallel to the z-axis and whose cross section by the plane $z = 0$ is bounded by the curve $(x^2 + y^2)^2 = a^2(x^2 - y^2)$.

13.9. Find the area of the surface $y = x \tan z$ $(|z| < \frac{1}{2}\pi)$, which lies inside the cylinder $x^2 + y^2 = 1$.

13.10. The paraboloid $z + 1 = x^2 + 4y^2$ meets the sphere $x^2 + y^2 + z^2 = 1$ in the curve C. Find the area of the smaller of the two regions into which the curve C divides the surface of the sphere.

13.11. A cone is defined by the equation $\phi = f(\theta)$. Show that the surface area of a portion of the cone is

$$\iint \sqrt{[1 + \{\sin \theta f'(\theta)\}^2]} \, r \, dr \, d\theta$$

over the appropriate field of integration in the r, θ plane. Hence find the area of the part of the cone $y^2z^2 = x^2(x^2 + y^2)$ in the first octant cut off by the surface of revolution $(x^2 + y^2 + z^2)^3 = z^2$.

SOLUTIONS

13.6. 2π.   **13.7.** $2\pi a^2 \sin^3\alpha$.   **13.8.** $2(\pi + 4 - 4\sqrt{2})a^2$.

13.9. $\pi\{\sqrt{2} + \log(1 + \sqrt{2})\}$.   **13.10.** π.

13.11. $\frac{1}{4}\{\sqrt{2} + \log(1 + \sqrt{2})\}$.

14. Surface Integrals

For a curve we distinguished between directions along the curve. For a surface we shall distinguish between sides of the surface. We are going to define the surface integral

$$\iint F(x, y, z) \, dx \, dy$$

over a portion S of one side of the surface $z = f(x, y)$. Let A be the projection of S on the plane $z = 0$. Let γ be the angle which the outward drawn normal to a particular side of the surface makes with the z-axis. For one side, the 'upper' side, $\cos \gamma$ is positive; for the other side, the 'lower' side, $\cos \gamma$ is negative. We define $\iint F(x, y, z) dx dy$ over the portion S of the side for which $\cos \gamma$ is positive as $+\iint\limits_{A} F[x, y, f(x, y)] dx dy$, and over the portion S of the other side as $-\iint\limits_{A} F[x, y, f(x, y)] dx dy$. If α, β are the angles which the outward drawn normal to a side of the surface make with the x, y-axes respectively, we define

$$\iint F(x, y, z) dy dz, \qquad \iint F(x, y, z) dz dx$$

as

$$\pm \iint F[x(y, z), y, z] dy dz, \qquad \pm \iint F[x, y(x, z), z] dz dx$$

over the projections on the planes $x = 0$, $y = 0$, taking $\pm$ signs according as $\cos \alpha$, $\cos \beta$ are positive or negative.

We define the surface integral

$$\iint F(x, y, z) dS$$

over the portion S of the surface $z = f(x, y)$ as

$$\iint\limits_{A} F[x, y, f(x, y)] \frac{dx dy}{|\cos \gamma|} = \iint\limits_{A} F[x, y, f(x, y)] \sqrt{(1 + f_x^2 + f_y^2)} dx dy,$$

where γ is the angle which either of the outward-drawn normals makes with the z-axis. It is clear from this definition that this surface integral does not depend on a particular side of the surface. We often write

$$dS = \sqrt{(1 + f_x^2 + f_y^2)} dx dy.$$

The following theorem is *Gauss's theorem* in three dimensions (the *divergence theorem* in vector analysis):

If P, Q, R are functions of x, y, z defined inside and on the boundary surface S of a solid figure K, then

$$\iiint_K \left(\frac{\partial P}{\partial x} + \frac{\partial Q}{\partial y} + \frac{\partial R}{\partial z} \right) dxdydz = \iint_S (Pdydz + Qdzdx + Rdxdy),$$

where the surface integral is taken over the outside of the surface S.

WORKED EXAMPLES

14.1. *Evaluate* $\iint z\, dxdy$ *over the outside surface of the sphere* $x^2 + y^2 + z^2 = 1$.

For the upper half of the sphere the integral is
$+\iint \sqrt{(1 - x^2 - y^2)}\, dxdy$ over the circle $x^2 + y^2 = 1$, that is

$$\int_0^{2\pi} d\theta \int_0^1 \sqrt{(1 - r^2)}\, r\, dr = 2\pi \left[-\tfrac{1}{3}(1 - r^2)^{3/2} \right]_0^1 = \frac{2\pi}{3}.$$

For the lower half of the sphere, the integral is $-\iint -\sqrt{(1 - x^2 - y^2)}\, dxdy$ over the same circle with, as above, value $\tfrac{2}{3}\pi$. Thus the total value of the integral is $\tfrac{4}{3}\pi$.

Alternately, we can use Gauss's theorem, taking $P = 0$, $Q = 0$, $R = z$ and we see that the surface integral is $\iiint 1\, dxdydz$ throughout the sphere, i.e., it is equal to $\tfrac{4}{3}\pi$, the volume of the sphere.

This is an example of the general result that a volume K is given by each of the surface integrals

$$\iint x\, dydz, \qquad \iint y\, dzdx, \qquad \iint z\, dxdy$$

taken over the outside of the boundary surface of K.

14.2. *Evaluate* $\iint \frac{x^2}{z}\, dS$ *over the portion of the surface of the para-boloid* $2z = x^2 + y^2$ *intercepted by the cylinder* $x^2 + y^2 = 1$.

On the paraboloid $p = \partial z / \partial x = x$, $q = \partial z / \partial y = y$ and hence

$$dS = \sqrt{(1 + x^2 + y^2)}\, dxdy.$$

Thus the surface integral is

$$\iint \frac{2x^2}{x^2+y^2} \sqrt{(1+x^2+y^2)}\,dxdy, \text{ over the circle } x^2+y^2 = 1,$$

$$= 2 \int_0^{2\pi} \cos^2 \theta \, d\theta \int_0^1 \sqrt{(1+r^2)} \, r \, dr, \text{ on changing to polars,}$$

$$= 2 \cdot 4 \frac{\pi}{4} \left[\tfrac{1}{3}(1+r^2)^{3/2} \right]_0^1 = \frac{2\pi}{3} (2\sqrt{2}-1).$$

14.3. *Prove that*

(*i*) $\quad \iiint\limits_K \nabla^2 u \, dxdydz = \iint\limits_S \dfrac{\partial u}{\partial n} \, dS;$

(*ii*) $\quad \iiint\limits_K (u\nabla^2 v - v\nabla^2 u)\,dxdydz = \iint\limits_S \left(u \dfrac{\partial v}{\partial n} - v \dfrac{\partial u}{\partial n} \right) dS;$

where $\nabla^2 = \partial^2/\partial x^2 + \partial^2/\partial y^2 + \partial^2/\partial z^2$ *and* $\partial/\partial n$ *is differentiation in the direction of the outward normal to the outside of S, the boundary surface of K;*

(*iii*) *if* $\nabla^2 u = 0$ *in K and v has a constant value on S, then*

$$\iiint\limits_K u\nabla^2 v \, dxdydz = \iint\limits_S u \dfrac{\partial v}{\partial n} \, dS;$$

(*iv*) *verify that the result of* (*iii*) *holds when* $u = r^{-1}$, $v = r^2 - r(a+b)$ *and K is the volume between the spheres* $r = a$, $r = b$, *where* $r^2 = x^2+y^2+z^2$ *and* $b > a > 0$.

(*i*) This is the extension to three dimensions of **12.3**.

Here, in Gauss's theorem, we let $P = \partial u/\partial x$, $Q = \partial u/\partial y$, $R = \partial u/\partial z$ and obtain

$$\iiint\limits_K \nabla^2 u \, dxdydz = \iint\limits_S \left(\frac{\partial u}{\partial x} \, dy \, dz + \frac{\partial u}{\partial y} \, dzdx + \frac{\partial u}{\partial z} \, dxdy \right)$$

over the outside of S. Let γ be the angle which the outward normal makes with the positive z-axis and consider

$$\iint_S \frac{\partial u}{\partial z}\, dxdy = \iint_S \frac{\partial u}{\partial z} \cos \gamma\, dS, \text{ over the upper portion of } S,$$

$$+ \iint_S \frac{\partial u}{\partial z} \cos \gamma\, dS, \text{ over the lower portion of } S,$$

$$= \iint_S \frac{\partial u}{\partial z} \cos \gamma\, dS, \text{ over the whole of } S.$$

since $\cos \gamma$ is positive on the upper portion and negative on the lower.

The other two terms in the surface integral give

$$\iint_S \frac{\partial u}{\partial x} \cos \alpha\, dS, \quad \iint_S \frac{\partial u}{\partial y} \cos \beta\, dS,$$ where α, β are the angles made by the outward normal with the positive x, y-axes. Hence

$$\iiint_K \nabla^2 u\, dxdydz = \iint_S \left(\frac{\partial u}{\partial x} \cos \alpha + \frac{\partial u}{\partial y} \cos \beta + \frac{\partial u}{\partial z} \cos \gamma \right) dS$$

$$= \iint_S \frac{\partial u}{\partial n}\, dS, \quad \text{by (4.6).}$$

(*ii*) This is the extension to three dimensions of **12.6.** Here we take in Gauss's theorem $P = u\partial v/\partial x - v\partial u/\partial x$, $Q = u\partial v/\partial y - v\partial u/\partial y$, $R = u\partial v/\partial z - v\partial u/\partial z$. The proof is as before and may be left to the reader.

(*iii*) In this case, since $\nabla^2 u = 0$ in K and v is constant on S, (*ii*) becomes

$$\iiint_K u\nabla^2 v\, dxdydz = \iint_S u \frac{\partial v}{\partial n}\, dS - v \iint_S \frac{\partial u}{\partial n}\, dS,$$

$$= \iint_S u \frac{\partial v}{\partial n}\, dS - v \iiint_K \nabla^2 u\, dxdydz, \quad \text{by (i),}$$

$$= \iint_S u \frac{\partial v}{\partial n}\, dS, \quad \text{since } \nabla^2 u = 0 \text{ in } K.$$

(*iv*) If $u(x, y, z) = f(r)$ where $r^2 = x^2 + y^2 + z^2$,

$$\frac{\partial u}{\partial x} = \frac{x}{r} f'(r), \quad \frac{\partial^2 u}{\partial x^2} = \left(\frac{1}{r} - \frac{x^2}{r^3}\right) f'(r) + \frac{x^2}{r^2} f''(r)$$

and similarly for derivatives with respect to y and z. Hence

$$\nabla^2 u = f'(r) \left(\frac{3}{r} - \frac{x^2 + y^2 + z^2}{r^3}\right) + f''(r) \left(\frac{x^2 + y^2 + z^2}{r^2}\right)$$

$$= \frac{2}{r} f'(r) + f''(r).$$

We now verify that $u = r^{-1}$ satisfies $\nabla^2 u = 0$; this is so since

$$\frac{2}{r} \frac{d}{dr} (r^{-1}) + \frac{d^2}{dr^2} (r^{-1}) = -\frac{2}{r^3} + \frac{2}{r^3} = 0.$$

Now since $v = r^2 - r(a+b)$, $\nabla^2 v = 6 - \frac{2}{r}(a+b)$.

Hence

$$\iiint u \nabla^2 v \, dx dy dz = \iiint_K \left(\frac{6}{r} - \frac{2}{r^2}(a+b)\right) r^2 \sin\theta \, dr d\theta d\phi$$

$$= \int_0^{2\pi} d\phi \int_0^\pi \sin\theta \, d\phi \int_a^b [6r - 2(a+b)] \, dr$$

$$= 4\pi \left[3r^2 - 2(a+b)r\right]_a^b = 4\pi(b^2 - a^2).$$

On $r = b$, $v = b^2 - b(a+b) = -ab$ and on $r = a$, $v = a^2 - a(a+b)$
$= -ab$, so that v is constant on the boundary surface S of K. On
$r = b$, $\partial/\partial n = \partial/\partial r$ and hence $\iint u \frac{\partial v}{\partial n} \, dS$ on $r = b$ is

$$\iint \frac{1}{r} (2r - (a+b)) \, dS = \iint \left(2 - \frac{a+b}{r}\right) dS$$

$$= \iint \left(2 - \frac{a+b}{b}\right) dS, \text{ since } r = b,$$

$$= 4\pi b^2 \left(2 - \frac{a+b}{b}\right) = 4\pi(b^2 - ab),$$

since the area of the sphere is $4\pi b^2$. On $r = a$, $\partial/\partial n = -\partial/\partial r$ and
hence as above,

$$\iint u \frac{\partial v}{\partial n}\, dS \text{ on } r = a \text{ is } -4\pi(a^2 - ab).$$

Thus over the whole of S,

$$\iint u \frac{\partial v}{\partial n}\, dS = 4\pi(b^2 - a^2).$$

14.4. *Show that if p is the length of the perpendicular from the origin to the tangent plane at a point on the hyperboloid $x^2/a^2 + y^2/b^2 - z^2/c^2 = 1$ $(a, b, c > 0)$, then*

$$\iint p\, dS = \tfrac{3}{2}\pi abc,$$

where the integral is taken over the surface of the hyperboloid between the planes $z = 0$ and $z = \tfrac{3}{4}c$.

The tangent plane at (x, y, z) on the hyperboloid is

$$(X-x)x/a^2 + (Y-y)y/b^2 - (Z-z)z/c^2 = 0 \quad \text{(see (4.3))}$$

i.e., $Xx/a^2 + Yy/b^2 - Zz/c^2 = 1$. Hence

$$p = \frac{1}{\sqrt{\{(x^2/a^4) + (y^2/b^4) + (z^2/c^4)\}}}.$$

Now $\partial z/\partial x = xc^2/za^2$ $\partial z/\partial y = yc^2/zb^2$ so that

$$\sqrt{\left\{1 + \left(\frac{\partial z}{\partial x}\right)^2 + \left(\frac{\partial z}{\partial y}\right)^2\right\}} = \sqrt{\left(1 + \frac{x^2c^4}{z^2a^4} + \frac{y^2c^4}{z^2b^4}\right)}$$

$$= \frac{c^2}{z}\sqrt{\left(\frac{x^2}{a^4} + \frac{y^2}{b^4} + \frac{z^2}{c^4}\right)}.$$

Hence

$$\iint p\, dS = \iint p \sqrt{\left\{1 + \left(\frac{\partial z}{\partial x}\right)^2 + \left(\frac{\partial z}{\partial y}\right)^2\right\}}\, dx\, dy, \quad \text{see (13.1)}$$

$$= c \iint \frac{dx\, dy}{\sqrt{\left(\frac{x^2}{a^2} + \frac{y^2}{b^2} - 1\right)}}$$

over the area between the ellipses,

$$x^2/a^2 + y^2/b^2 = 25/16 \quad \text{and} \quad x^2/a^2 + y^2/b^2 = 1.$$

We now let $x = a\xi$, $y = b\eta$, and we have

$$\iint p \, dS = abc \iint \frac{d\xi d\eta}{\sqrt{(\xi^2 + \eta^2 - 1)}}$$

over the area between the circles $\xi^2 + \eta^2 = 25/16$ and $\xi^2 + \eta^2 = 1$, i.e.

$$\iint p \, dS = 4abc \int_0^{\pi/2} d\theta \int_1^{5/4} \frac{r \, dr}{\sqrt{(r^2 - 1)}}$$

$$= 2\pi abc \left[(r^2 - 1)^{1/2} \right]_1^{5/4} = \tfrac{3}{2}\pi abc.$$

14.5. *Show that the mean value of the square of the distance of points on the surface of a sphere of radius a from a tangent plane to the sphere is $\tfrac{4}{3}a^2$.*

The mean value of $f(x, y, z)$ over a surface is defined to be $\iint f(x, y, z) \, dS$ over the surface divided by the area of the surface.

In our problem let the sphere be $x^2 + y^2 + z^2 = 2az$ and the tangent plane $z = 0$ so that the required mean value is $\dfrac{1}{4\pi a^2} \iint z^2 dS$ over the sphere. In spherical polar coordinates the sphere is $r = 2a \cos \theta$. Hence parametric equations of the sphere are

$$x = a \sin 2\theta \cos \phi, \quad y = a \sin 2\theta \sin \phi, \quad z = 2a \cos^2 \theta$$

taking θ and ϕ as the parameters. Now

$$dS = \sqrt{\left[\left\{ \frac{\partial(y, z)}{\partial(\theta, \phi)} \right\}^2 + \left\{ \frac{\partial(z, x)}{\partial(\theta, \phi)} \right\}^2 + \left\{ \frac{\partial(x, y)}{\partial(\theta, \phi)} \right\}^2 \right]} \, d\theta \, d\phi,$$

$$= \sqrt{(4a^4 \sin^4 2\theta \cos^2 \phi + 4a^4 \sin^4 2\theta \sin^2 \phi}$$

$$+ 4a^4 \cos^2 2\theta \sin^2 2\theta) d\theta d\phi$$

$$= 2a^2 \sin 2\theta d\theta d\phi.$$

Hence the mean value is

$$\frac{1}{4\pi a^2} \int_0^{2\pi} d\phi \int_0^{\pi/2} 16a^4 \cos^5 \theta \sin \theta \, d\theta = \frac{1}{4\pi a^2} (2\pi) \left(\frac{16a^4}{6} \right)$$

$$= \frac{4a^2}{3}.$$

14.6. *If C is the point* $(0, 0, c)$ *and P is the point* (x, y, z) *on the surface S of the sphere* $x^2 + y^2 + z^2 = a^2$, *show that*

$$\iint_S \frac{c-z}{CP^3}\, dS = \begin{cases} \dfrac{4\pi a^2}{c^2}, & c > a, \\[2mm] 2\pi, & c = a, \\[1mm] 0, & c < a. \end{cases}$$

The sphere has parametric equations

$$x = a \sin\theta \cos\phi, \quad y = a \sin\theta \sin\phi, \quad z = a \cos\theta.$$

An easy calculation shows that

$$dS = a^2 \sin\theta\, d\theta d\phi$$

(this can also be seen geometrically, since when θ takes the increment $\Delta\theta$ and ϕ the increment $\Delta\phi$, the area of the corresponding increment of surface of the sphere is approximately $(a\Delta\theta)(a \sin\theta\Delta\phi)$). The surface integral is

$$\iint_S \frac{(c - a\cos\theta)\, a^2 \sin\theta\, d\theta d\phi}{(c^2 + a^2 - 2ca\cos\theta)^{3/2}}$$

$$= \int_0^{2\pi} d\phi \int_0^{\pi} \frac{(c - a\cos\theta)\, a^2 \sin\theta\, d\theta}{(c^2 + a^2 - 2ca\cos\theta)^{3/2}}$$

$$= 2\pi a^2 \int_{-1}^{1} \frac{(c - au)\, du}{(c^2 + a^2 - 2cau)^{3/2}}, \quad \text{where } u = \cos\theta,$$

$$= 2\pi a^2 \int_{-1}^{1} \left\{ \frac{1}{2c} (c^2 + a^2 - 2cau)^{-1/2} \right.$$

$$\left. + \left(\frac{c}{2} - \frac{a^2}{2c}\right)(c^2 + a^2 - 2cau)^{-3/2} \right\} du, \quad c \neq a,$$

$$= 2\pi a^2 \left\{ -\frac{1}{2c^2 a}\left[(c^2 + a^2 - 2cau)^{1/2}\right]_{-1}^{1} \right.$$

$$\left. + \frac{1}{ca}\left(\frac{c}{2} - \frac{a^2}{2c}\right)\left[(c^2 + a^2 - 2cau)^{-1/2}\right]_{-1}^{1} \right\}.$$

If $c > a$, this is

$$-\frac{\pi a}{c^2}\left\{(c-a)-(c+a)\right\}+\frac{\pi a^2(c^2-a^2)}{c^2 a}\left(\frac{1}{c-a}-\frac{1}{c+a}\right)$$

$$=\frac{2\pi a^2}{c^2}+\pi a^2\left(\frac{1}{c^2 a}\right)2a=\frac{4\pi a^2}{c^2}.$$

If $c < a$, the value is

$$-\frac{\pi a}{c^2}(a-c-c-a)+\frac{\pi a}{c^2}(c^2-a^2)\left(\frac{1}{a-c}-\frac{1}{a+c}\right)$$

$$=+\frac{2\pi a}{c}-\frac{2\pi a}{c}=0.$$

If $c = a$, the integral is

$$2\pi\int_{-1}^{1}\frac{1}{2\sqrt{2}}(1-u)^{-1/2}\,du=\frac{\pi}{\sqrt{2}}\left[-2(1-u)^{1/2}\right]_{-1}^{1}=2\pi.$$

ADDITIONAL EXAMPLES

14.7. The surface integral $\iint x^2(1-z^2)\,dxdy$ is taken over the outside surface of the solid hemisphere bounded by the upper half of the sphere $x^2+y^2+z^2=1$ and the plane $z=0$. Find the value of the surface integral (i) directly; (ii) using Gauss's theorem.

Evaluate $\iint x^2(1-z^2)\,dS$ taken over the same surface.

14.8. Evaluate

$$\iint \frac{1}{xyz}(yz\,dydz+zx\,dzdx+xy\,dxdy)$$

over the outside of the surface of the sphere $x^2+y^2+z^2=1$.

14.9. Evaluate

$$\iint_S (x^3\,dydz+x^2y\,dzdx+x^2z\,dxdy),$$

where S is the entire surface of the cylinder $x^2+y^2=a^2, 0\leqslant z\leqslant b$.

14.10. If C is a point distant c $(c > a)$ from the centre of a sphere of radius a, find the mean value of the inverse of the square of the distance from C to points on the surface of the sphere.

14.11. Show that the mean value of $x^2y^2z^2$ for points on the surface of the sphere $x^2+y^2+z^2 = a^2$ is $a^6/105$.

If P is the point $(0, 0, c)$ $(c > a)$ and Q is any point on the surface of the sphere, prove that the mean value of $1/PQ^3$ is $1/c(c^2-a^2)$. Find its mean value when $c < a$.

14.12. If $f(x, y, z) = 0$ defines z as a single-valued function of x and y and if the portion A of the surface $f(x, y, z) = 0$ has projection A' on the plane $z = 0$, show that

$$\iint_A p \, dS = \iint_{A'} \left| \frac{\partial F}{\partial t} \middle/ \frac{\partial F}{\partial z} \right|_{t=1} dxdy,$$

where p is the length of the perpendicular from the origin to the tangent plane at the point (x, y, z) and

$$F(x, y, z, t) \equiv t^n f(x/t, y/t, z/t).$$

Show that this integral has the value $\pi\sqrt{3}$ when taken over the part of the surface of the hyperboloid

$$x^2+xy+y^2-z^2 = 1$$

which lies between the planes $z = 0$ and $z = 3/4$.

SOLUTIONS

14.7. $-\pi/12,\ 23\pi/60.$ **14.8.** $12\pi.$ **14.9.** $\frac{5}{4}\pi\, a^4b.$

14.10. $\dfrac{1}{2ac} \log \dfrac{c+a}{c-a}.$ **14.11.** $1/a\,(a^2-c^2).$

15. Kelvin's (or Stokes's) theorem

Let S be a portion of a surface which is bounded by C, a curve in space. Associated with each side of S we define a positive direction

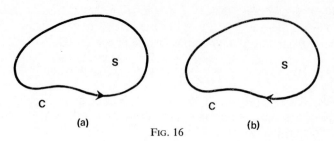

FIG. 16

round C in such a way that a person going round C in the positive direction for a particular side has that side on his left. In Fig. 16 (a) the arrowhead denotes the positive direction round C for the 'upper' side of S and in Fig. 16 (b) the arrowhead denotes the positive direction for the 'lower' side.

Let S have equation $z = f(x, y)$ and let F be a function of x, y, z. Let C_1 be the plane curve which is the projection of the curve C in the plane $z = 0$. As a point moves round C in a certain direction its projection moves round C_1 in a corresponding direction. We define the curvilinear integral $\int_C F(x, y, z)dx$ in a certain direction round C as $\int_{C_1} F[x, y, f(x, y)]dx$ in the corresponding direction round C_1. We define similarly $\int_C Fdy$ and $\int_C Fdz$.

We now state Kelvin's theorem. (This theorem, long attributed to Stokes, is now known to have been first proved by William Thomson, Lord Kelvin.)

If F, G, H are functions of x, y, z, then

$$\int_C Fdx + Gdy + Hdz$$

$$= \iint_S \left\{ l\left(\frac{\partial H}{\partial y} - \frac{\partial G}{\partial z}\right) + m\left(\frac{\partial F}{\partial z} - \frac{\partial H}{\partial x}\right) + n\left(\frac{\partial G}{\partial x} - \frac{\partial F}{\partial y}\right) \right\} dS,$$

where C is the boundary curve of S, the curvilinear integral on the left is taken in the positive direction associated with a certain side of S and (l, m, n) are the direction cosines of the outward normal to that side of S.

WORKED EXAMPLES

15.1. *Evaluate $\int xyz\,dx$ round the ellipse which is the intersection of the cylinder $x^2+y^2 = 1$ by the plane $x+z = 1$, the direction of integration being the positive one associated with the 'upper' side of the ellipse.*

The projection of the ellipse on the plane $z = 0$ is the circle $x^2+y^2 = 1$ and we have to integrate $xy(1-x)$ round this circle in the positive direction. This integral is

$$\int_1^{-1} x\sqrt{(1-x^2)}(1-x)\,dx + \int_{-1}^1 x\{-\sqrt{(1-x^2)}\}(1-x)\,dx$$

$$= -2\int_{-1}^1 x\sqrt{(1-x^2)}(1-x)\,dx = 4\int_0^1 x^2\sqrt{(1-x^2)}\,dx$$

$$= 4\int_0^{\pi/2} \sin^2 u\cos^2 u\,du, \text{ where } x = \sin u,$$

$$= 4\cdot\frac{1}{4.2}\frac{\pi}{2} = \frac{\pi}{4}.$$

15.2. *Verify that Kelvin's theorem is true when S is the part of the*

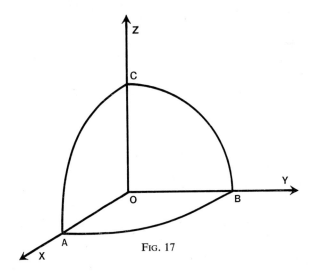

FIG. 17

surface of the sphere $x^2+y^2+z^2 = 1$ *in the first octant and* $F = z$, $G = x$, $H = y$.

Let us choose the 'upper' side of S so that we have to take the line integral $\int (zdx+xdy+ydz)$ round the boundary curve in the direction $ABCA$. On AB this integral is $\int_0^1 \sqrt{(1-y^2)}\,dy$, since $z = 0$ and $\int ydz = 0$. The value for this part of the line integral is thus $\frac{1}{4}\pi$. On BC the integral is $\int_0^1 \sqrt{(1-z^2)}\,dz = \frac{1}{4}\pi$ and on CA it is $\int_0^1 \sqrt{(1-x^2)}\,dx = \frac{1}{4}\pi$.

Hence the complete line integral is $\frac{3}{4}\pi$.

We have now to calculate

$$\iint_S l\left(\frac{\partial H}{\partial y}-\frac{\partial G}{\partial z}\right)+m\left(\frac{\partial F}{\partial z}-\frac{\partial H}{\partial x}\right)+n\left(\frac{\partial G}{\partial x}-\frac{\partial F}{\partial y}\right)dS$$

$$= \iint_S (l+m+n)\,dS.$$

For the 'upper' side of S, $(l, m, n) = (x, y, z)$, where $z = \sqrt{(1-x^2-y^2)}$ and $dS = dxdy/\sqrt{(1-x^2-y^2)}$. Hence the surface integral is $\iint \left(\frac{x+y}{\sqrt{(1-x^2-y^2)}}+1\right) dxdy$ over the first quadrant of the circle $x^2+y^2 = 1$,

$$= \int_0^{\pi/2} (\cos\theta+\sin\theta)\,d\theta \int_0^1 \frac{r^2}{\sqrt{(1-r^2)}}\,dr+\frac{\pi}{4}$$

$$= 2\int_0^{\pi/2} \sin^2 u\,du+\frac{\pi}{4}, \text{ where } r = \sin u,$$

$$= 2\frac{\pi}{4}+\frac{\pi}{4} = \frac{3\pi}{4}.$$

If we had chosen the 'lower' side of S, we would have obtained $-\frac{3}{4}\pi$ for both the line and surface integrals.

15.3. *Evaluate* $\int x \, dz$ *round the curve of intersection of the hemisphere* $z = \sqrt{(1-x^2-y^2)}$ *and the cylinder* $x^2+y^2 = y$ *where the direction round the curve is from the point* $(0, 0, 1)$ *into the first octant.*

We use Kelvin's theorem to express the line integral as a surface integral over the portion of the surface of the hemisphere cut off by the cylinder. We have to take the 'upper' side of the surface to correspond with the given direction along the curve. In the theorem we take $F = G = 0$, $H = x$ so that the surface integral is

$$\iint (-m) \, dS$$

where, as in the last example, $m = y$, $dS = dxdy/\sqrt{(1-x^2-y^2)}$. Hence the surface integral becomes

$$\iint \frac{-y \, dxdy}{\sqrt{(1-x^2-y^2)}}$$

over the circle $x^2+y^2 = y$, i.e.,

$$-2 \int_0^{\pi/2} \sin \theta \, d\theta \int_0^{\sin \theta} \frac{r^2}{\sqrt{(1-r^2)}} \, dr$$

$$= -2 \int_0^{\pi/2} \sin \theta \, d\theta \int_0^{\theta} \sin^2 u \, du, \text{ where } r = \sin u,$$

$$= - \int_0^{\pi/2} (\theta \sin \theta - \sin^2 \theta \cos \theta) d\theta = -1+\tfrac{1}{3} = -\tfrac{2}{3}.$$

ADDITIONAL EXAMPLES

15.4. S is the portion of the plane $y+z = 1$ cut off by the cylinder $x^2+y^2 = 1$ and C is the boundary of S. Evaluate

$$\int_C (y-z)dx+(z-x)dy+(x-y)dz,$$

where the integral is taken round C in the positive direction associated with the 'upper' side of S.

15.5. Verify Kelvin's theorem when $F = y^2$, $G = z^2$, $H = x^2$ and S is the portion of the plane $x+y+z = 1$ which lies in the first octant.

SOLUTION

15.4. -4π.

Chapter V

VECTOR CALCULUS

16. Grad, div and curl

All the work in this chapter concerns problems in three-dimensional space and the vectors we consider have three components. Suppose that at each point P of a domain D we define a unique scalar function ϕ; then we denote the fact that ϕ depends on the coordinates (x, y, z) of P by writing it as $\phi(\mathbf{r})$, where $\mathbf{r}$ is the *position vector* of P with components (x, y, z). The vector $\mathbf{r}$ can also be written as

$$x\mathbf{i}+y\mathbf{j}+z\mathbf{k},$$

where $\mathbf{i}, \mathbf{j}, \mathbf{k}$ are respectively unit vectors in the x, y, z directions (see *Solving Problems in Vector Algebra*, §1). ϕ is said to define a scalar field in D.

The electrostatic potential is a scalar field.

If at each point of D there is defined a vector $\mathbf{v}$, we say that $\mathbf{v}(\mathbf{r})$ defines a *vector field* in D; $\mathbf{v}$ will have x, y, z components v_1, v_2, v_3, say, where each of these is a scalar function. The electric force is an example of a vector field.

From a given scalar field ϕ we can derive a vector field, the *gradient vector field* or grad ϕ, by defining it as the vector field with components

$$\left(\frac{\partial \phi}{\partial x}, \frac{\partial \phi}{\partial y}, \frac{\partial \phi}{\partial z}\right). \qquad (16.1)$$

It can be easily shown that the derivative of ϕ in the direction defined by the unit vector $\mathbf{n}$ (see §4) is given by the scalar product $\mathbf{n} \cdot \text{grad } \phi$ (see *Solving Problems in Vector Algebra*, §8).

Suppose that a vector field $\mathbf{v}$ is defined in a domain D and that P is an interior point of D. In the neighbourhood of P we construct a small region R with surface S whose volume is τ; then

$$\frac{1}{\tau} \iint_S (\mathbf{v} \cdot \mathbf{n}) \, dS,$$

where **n** is the unit vector in the outward direction normal to S at a typical point of S, will be a scalar representing the flux of **v** out of R per unit volume. If the limit

$$\lim_{\tau \to 0} \frac{1}{\tau} \iint\limits_{S} (\mathbf{v} \cdot \mathbf{n}) \, dS,$$

taken so that R always contains P, exists, we say that it defines the *divergence of* **v** at P, written div **v**. Thus div **v** is a scalar field derived from the vector field **v**. It can be shown that, if **v** is defined as $v_1\mathbf{i} + v_2\mathbf{j} + v_3\mathbf{k}$, then

$$\text{div } \mathbf{v} = \frac{\partial v_1}{\partial x} + \frac{\partial v_2}{\partial y} + \frac{\partial v_3}{\partial z}. \qquad (16.2)$$

The Laplace operator ∇^2 (see **2.2**) as applied to the scalar field ϕ, is defined as

$$\nabla^2 \phi = \text{div grad } \phi = \frac{\partial^2 \phi}{\partial x^2} + \frac{\partial^2 \phi}{\partial y^2} + \frac{\partial^2 \phi}{\partial z^2}. \qquad (16.3)$$

The Laplace operator is also used as applied to a vector field **v** in the following way; by $\nabla^2 \mathbf{v}$ we mean the vector field with components

$$(\nabla^2 v_1, \nabla^2 v_2, \nabla^2 v_3), \qquad (16.4)$$

where v_1, v_2, v_3 are the x, y, z components of **v** respectively.

We can derive from a given vector field **v** another vector field, called the *curl* of **v**, in the following way. At a point P in the region in which **v** is defined, we construct a small surface S bounded by a closed curve C. The normal at P to a particular side of S is described by the unit vector **n** and we choose the positive direction round C associated with this side. (See §15.) Let v_s be the component of **v** along the tangent to C taken in this positive direction. We then form the limit

$$\lim_{S \to 0} \frac{1}{S} \int\limits_{C} v_s ds,$$

taken so that C shrinks to the point P as $S \to 0$. The value of the limit (if it exists) is defined to be the component of the vector curl **v** in the direction **n**. It can be shown that, if $\mathbf{v} = v_1\mathbf{i} + v_2\mathbf{j} + v_3\mathbf{k}$, then

$$\operatorname{curl} \mathbf{v} = \left(\frac{\partial v_3}{\partial y} - \frac{\partial v_2}{\partial z}, \; \frac{\partial v_1}{\partial z} - \frac{\partial v_3}{\partial x}, \; \frac{\partial v_2}{\partial x} - \frac{\partial v_1}{\partial y} \right). \qquad (16.5)$$

The concepts of grad, div and curl can be expressed in terms of the *del operator* $\mathbf{V}$, which is the vector operator

$$\mathbf{V} = \left(\frac{\partial}{\partial x}, \; \frac{\partial}{\partial y}, \; \frac{\partial}{\partial z} \right). \qquad (16.6)$$

Thus if ϕ is a scalar field and $\mathbf{v}$ a vector field

$$\operatorname{grad} \phi = \mathbf{V}\phi,$$
$$\operatorname{div} \mathbf{v} = \mathbf{V} . \mathbf{v}, \qquad (16.7)$$
$$\operatorname{curl} \mathbf{v} = \mathbf{V} \times \mathbf{v}.$$

Vector Identities

Throughout ϕ denotes a scalar field and $\mathbf{u}$ and $\mathbf{v}$ vector fields.

$$\operatorname{div} (\phi\mathbf{u}) = \phi \operatorname{div} \mathbf{u} + \mathbf{u} . \operatorname{grad} \phi, \qquad (16.8)$$

$$\operatorname{curl} \operatorname{grad} \phi = \mathbf{0}, \qquad (16.9)$$

where $\mathbf{0}$ denotes the zero vector $(0, 0, 0)$.

$$\operatorname{div} \operatorname{curl} \mathbf{u} = 0, \qquad (16.10)$$

where 0 here denotes the scalar zero.

$$\operatorname{div} (\mathbf{u} \times \mathbf{v}) = \mathbf{v} . \operatorname{curl} \mathbf{u} - \mathbf{u} . \operatorname{curl} \mathbf{v}, \qquad (16.11)$$

$$\operatorname{curl} \operatorname{curl} \mathbf{u} = \operatorname{grad} \operatorname{div} \mathbf{u} - \nabla^2 \mathbf{u}, \qquad (16.12)$$

$$\operatorname{curl} (\phi\mathbf{u}) = \phi \operatorname{curl} \mathbf{u} - \mathbf{u} \times \operatorname{grad} \phi, \qquad (16.13)$$

$$\operatorname{curl} (\mathbf{u} \times \mathbf{v}) = \mathbf{u} \operatorname{div} \mathbf{v} - \mathbf{v} \operatorname{div} \mathbf{u}$$
$$+ (\mathbf{v} . \operatorname{grad})\mathbf{u} - (\mathbf{u} . \operatorname{grad})\mathbf{v}. \qquad (16.14)$$

This notation requires some explanation. By $(\mathbf{u} . \operatorname{grad}) \phi$ we mean

$$u_1 \frac{\partial \phi}{\partial x} + u_2 \frac{\partial \phi}{\partial y} + u_3 \frac{\partial \phi}{\partial z}$$

where $\mathbf{u} = (u_1, u_2, u_3)$, and by $(\mathbf{u} . \operatorname{grad})\mathbf{v}$ we mean the vector

$$((\mathbf{u} . \operatorname{grad})v_1, \; (\mathbf{u} . \operatorname{grad})v_2, \; (\mathbf{u} . \operatorname{grad})v_3),$$

where $\mathbf{v} = (v_1, v_2, v_3)$.

If a vector field **u** is such that curl **u** = 0, the field is said to be *irrotational* or *conservative* and there exists a scalar field V such that **u** = grad V. Such a field is called the scalar potential field associated with **u**.

It may happen in physical problems that the components (u_1, u_2, u_3) of a vector field **u** are functions of time t. In that case

$$\frac{d\mathbf{u}}{dt} = \frac{du_1}{dt}\mathbf{i} + \frac{du_2}{dt}\mathbf{j} + \frac{du_3}{dt}\mathbf{k}. \qquad (16.15)$$

WORKED EXAMPLES

16.1. *Express in terms of the del operator the identities* (16.8)–(16.14).

We use the notation given in (16.7) to get

(16.8) $\qquad \mathbf{V} \cdot (\phi \mathbf{u}) = \phi \mathbf{V} \cdot \mathbf{u} + \mathbf{u} \cdot \mathbf{V}\phi.$

(16.9) $\qquad \mathbf{V} \times \mathbf{V}\phi = \mathbf{0}.$

(16.10) $\qquad \mathbf{V} \cdot (\mathbf{V} \times \mathbf{u}) = 0.$

(16.11) $\qquad \mathbf{V} \cdot (\mathbf{u} \times \mathbf{v}) = \mathbf{v} \cdot (\mathbf{V} \times \mathbf{u}) - \mathbf{u} \cdot (\mathbf{V} \times \mathbf{v}).$

(16.12) $\qquad \mathbf{V} \times (\mathbf{V} \times \mathbf{u}) = \mathbf{V}(\mathbf{V} \cdot \mathbf{u}) - \mathbf{V}^2 \mathbf{u}.$

(16.13) $\qquad \mathbf{V} \times (\phi \mathbf{u}) = \phi(\mathbf{V} \times \mathbf{u}) - \mathbf{u} \times \mathbf{V}\phi.$

(16.14) $\qquad \mathbf{V} \times (\mathbf{u} \times \mathbf{v}) = \mathbf{u}(\mathbf{V} \cdot \mathbf{v}) - \mathbf{v}(\mathbf{V} \cdot \mathbf{u}) + (\mathbf{v} \cdot \mathbf{V})\mathbf{u} - (\mathbf{u} \cdot \mathbf{V})\mathbf{v}.$

16.2. *Evaluate* (i) grad (r^{2n}); (ii) div $(r^2\mathbf{r})$, *where* **r** *is the position vector* (x, y, z) *and* $r = |\mathbf{r}|$.

(i) The x-component of grad r^{2n} is $\partial(r^{2n})/\partial x$, which is $2nr^{2n-1}$. $(x/r) = 2nr^{2n-2}x$, since $\partial r/\partial x = x/r$. The other two components are obtained similarly and we have

$$\text{grad } (r^{2n}) = 2nr^{2n-2}(x\mathbf{i} + y\mathbf{j} + z\mathbf{k}) = 2nr^{2n-2}\mathbf{r}.$$

(ii) div $(r^2\mathbf{r}) = \dfrac{\partial}{\partial x}(r^2 x) + \dfrac{\partial}{\partial y}(r^2 y) + \dfrac{\partial}{\partial z}(r^2 z),$

$$= r^2 + 2r\left(\frac{x}{r}\right)x + r^2 + 2r\left(\frac{y}{r}\right)y + r^2 + 2r\left(\frac{z}{r}\right)z,$$

$$= 3r^2 + 2(x^2 + y^2 + z^2) = 5r^2.$$

16.3. *If ψ is a scalar field, show that*

$$\mathbf{V}.(\mathbf{r} \times \mathbf{V}\psi) = 0,$$

where $\mathbf{r}$ is the position vector (x, y, z).
If $r^2 = x^2 + y^2 + z^2$, show that

$$\mathbf{V}.\{f(r)(\mathbf{u} \times \mathbf{r})\} = \mathbf{r}.(\mathbf{V} \times \mathbf{u})f(r).$$

We have

$$\mathbf{r} \times \mathbf{V}\psi = \left(y\frac{\partial \psi}{\partial z} - z\frac{\partial \psi}{\partial y}, \ z\frac{\partial \psi}{\partial x} - x\frac{\partial \psi}{\partial z}, \ x\frac{\partial \psi}{\partial y} - y\frac{\partial \psi}{\partial x}\right),$$

so that

$$\mathbf{V}.(\mathbf{r} \times \mathbf{V}\psi) = y\frac{\partial^2 \psi}{\partial x\partial z} - z\frac{\partial^2 \psi}{\partial x\partial y} + z\frac{\partial^2 \psi}{\partial y\partial x} - x\frac{\partial^2 \psi}{\partial y\partial z} + x\frac{\partial^2 \psi}{\partial z\partial y}$$

$$- y\frac{\partial^2 \psi}{\partial z\partial x}$$

$$= 0.$$

Using (*16.8*) we have

$$\mathbf{V}.\{f(r)(\mathbf{u} \times \mathbf{r})\} = f(r)\mathbf{V}.(\mathbf{u} \times \mathbf{r}) + (\mathbf{u} \times \mathbf{r}).\mathbf{V}f(r)$$

$$= f(r)\{\mathbf{r}.\mathbf{V} \times \mathbf{u} - \mathbf{u}.\mathbf{V} \times \mathbf{r}\} + (\mathbf{u} \times \mathbf{r}).\frac{1}{r}f'(r)\mathbf{r},$$

where we have used (*16.11*). Now $\mathbf{V} \times \mathbf{r}$ is clearly the zero vector and $(\mathbf{u} \times \mathbf{r}).\mathbf{r} = [\mathbf{u}, \mathbf{r}, \mathbf{r}]$ is the scalar zero (see *Vector Algebra*, §**13**). Hence result.

16.4. *Maxwell's equations for the electromagnetic field in vacuo may be written in the form*

$$c \operatorname{curl} \mathbf{M} = i\frac{\partial \mathbf{M}}{\partial t}, \quad \operatorname{div} \mathbf{M} = 0,$$

where $\mathbf{M} = \mathbf{E} + i\mathbf{H}$, $\mathbf{E}$ and $\mathbf{H}$ are the electric and magnetic field strengths respectively, $i = \sqrt{-1}$, and c the velocity of light. Show that the equations are satisfied by

$$\mathbf{M} = \frac{i}{c}\operatorname{curl}\left(\frac{\partial\mathbf{S}}{\partial t}\right) + \operatorname{curl}\operatorname{curl}\mathbf{S},$$

where **S** *is a vector function of position and time satisfying the wave equation*

$$c^2\nabla^2\mathbf{S} = \frac{\partial^2\mathbf{S}}{\partial t^2}.$$

Show that

$$\frac{\mathbf{k}}{r}\sin\omega t\sin\left(\frac{\omega r}{c}\right),$$

where **k** *is a constant unit vector, r is the distance from the origin and ω is a real constant, is a possible form of* **S**.

Maxwell's equations are

$$c\operatorname{curl}(\mathbf{E}+i\mathbf{H}) = i\,\frac{\partial(\mathbf{E}+i\mathbf{H})}{\partial t}, \;\; \operatorname{div}(\mathbf{E}+i\mathbf{H}) = 0$$

and on equating real and imaginary parts they give

$$c\operatorname{curl}\mathbf{E} = -\frac{\partial\mathbf{H}}{\partial t}, \;\; c\operatorname{curl}\mathbf{H} = \frac{\partial\mathbf{E}}{\partial t}, \;\; \operatorname{div}\mathbf{E} = 0, \;\; \operatorname{div}\mathbf{H} = 0.$$

We take

$$\mathbf{E} = \operatorname{curl}\operatorname{curl}\mathbf{S}, \;\; \mathbf{H} = \frac{1}{c}\operatorname{curl}\frac{\partial\mathbf{S}}{\partial t}, \;\; \text{where } c^2\nabla^2\mathbf{S} = \frac{\partial^2\mathbf{S}}{\partial t^2},$$

and show that these four equations are satisfied.

Now $c\operatorname{curl}\mathbf{H} = \operatorname{curl}\operatorname{curl}\dfrac{\partial\mathbf{S}}{\partial t} = \dfrac{\partial}{\partial t}\operatorname{curl}\operatorname{curl}\mathbf{S} = \dfrac{\partial\mathbf{E}}{\partial t}.$

Again

$$\begin{aligned}
c\operatorname{curl}\mathbf{E} &= c\operatorname{curl}\operatorname{curl}\operatorname{curl}\mathbf{S} \\
&= c\operatorname{curl}(\operatorname{grad}\operatorname{div}\mathbf{S} - \nabla^2\mathbf{S}), \;\; \text{by } (16.12), \\
&= -c\operatorname{curl}\nabla^2\mathbf{S}, \;\; \text{by } (16.9), \\
&= -\frac{1}{c}\operatorname{curl}\frac{\partial^2\mathbf{S}}{\partial t^2} = -\frac{\partial\mathbf{H}}{\partial t}.
\end{aligned}$$

By (16.10) the third and fourth equations are clearly satisfied.

To complete the problem, we now take

$$S = \frac{k}{r} \sin \omega t \sin \left(\frac{\omega r}{c}\right).$$

Using $\partial r/\partial x = x/r$, we get

$$\frac{\partial^2 S}{\partial t^2} = -\frac{\omega^2}{r} \, k \sin \omega t \sin \frac{\omega r}{c},$$

$$\frac{\partial S}{\partial x} = -\frac{x}{r^3} \, k \sin \omega t \sin \frac{\omega r}{c} + \frac{x\omega}{cr^2} \, k \sin \omega t \cos \frac{\omega r}{c},$$

and

$$\frac{\partial^2 S}{\partial x^2} = k \sin \omega t \left\{ \sin \frac{\omega r}{c} \left(-\frac{1}{r^3} + \frac{3x^2}{r^5} - \frac{x^2\omega^2}{r^3c^2} \right) \right.$$

$$\left. + \cos \frac{\omega r}{c} \left(-\frac{3x^2\omega}{r^4c} + \frac{\omega}{r^2c} \right) \right\},$$

with similar expressions for $\delta^2 S/\delta y^2$, $\delta^2 S/\delta z^2$. On adding we get

$$\nabla^2 S = -\frac{\omega^2}{rc^2} \, k \sin \omega t \sin \frac{\omega r}{c} = \frac{1}{c^2} \frac{\partial^2 S}{\partial t^2}.$$

16.5. *If* v *is a constant vector,* r *is the position vector* (x, y, z) $(|r| = r)$ *and*

$$\phi = \frac{v \cdot r}{r^3},$$

show that $\nabla \cdot (\nabla \phi) = 0$ *and that* $\nabla \phi = -\nabla \times A$, *where* $A = (v \times r)/r^3$.

Let $v = (v_1, v_2, v_3)$. Since $\phi = (v_1 x + v_2 y + v_3 z)/r^3$,

$$\frac{\partial \phi}{\partial x} = \frac{v_1}{r^3} - \frac{3(v_1 x + v_2 y + v_3 z)x}{r^5},$$

$$\frac{\partial^2 \phi}{\partial x^2} = \frac{-3v_1 x}{r^5} - \frac{6v_1 x + 3v_2 y + 3v_3 z}{r^5} + \frac{15x^2(v_1 x + v_2 y + v_3 z)}{r^7},$$

and similarly for the terms $\partial^2 \phi/\partial y^2$, $\partial^2 \phi/\partial z^2$. It follows that $\nabla \cdot \nabla \phi$ consists of a sum of three terms $u_1 v_1 + u_2 v_2 + u_3 v_3$, where

$$u_1 = -\frac{9x}{r^5} + \frac{15x^3}{r^7} - \frac{3x}{r^5} + \frac{15xy^2}{r^7} - \frac{3x}{r^5} + \frac{15xz^2}{r^7}$$

$$= \frac{x}{r^7} \{15(x^2 + y^2 + z^2) - 15r^2\} = 0$$

and similarly $u_2 = u_3 = 0$. Hence $\mathbf{V} \cdot \mathbf{V}\phi = 0$.
Now

$$\mathbf{A} = \frac{\mathbf{v} \times \mathbf{r}}{r^3} = \frac{1}{r^3}(v_2 z - v_3 y, \; v_3 x - v_1 z, \; v_1 y - v_2 x),$$

so that the x-component of $-\mathbf{V} \times \mathbf{A}$ is

$$\frac{\partial}{\partial z} \frac{1}{r^3}(v_3 x - v_1 z) - \frac{\partial}{\partial y} \frac{1}{r^3}(v_1 y - v_2 x)$$

$$= -\frac{2v_1}{r^3} - \frac{3}{r^5}\{z(v_3 x - v_1 z) - y(v_1 y - v_2 x)\}$$

$$= \frac{v_1}{r^3} - \frac{3(v_1 x + v_2 y + v_3 z)}{r^5} x,$$

which is the x-component of $\mathbf{V}\phi$. We have similar results for the other components and hence

$$\mathbf{V}\phi = -\mathbf{V} \times \mathbf{A}.$$

16.6. *If $\mathbf{r}$ is the position vector of a point in space and $r = |\mathbf{r}|$, show that for any scalar field $\psi(\mathbf{r})$,*

$$\operatorname{grad} \psi = \left(\frac{\partial \psi}{\partial x}\right)_{y,r} \mathbf{i} + \left(\frac{\partial \psi}{\partial y}\right)_{x,r} \mathbf{j} + \frac{1}{r}\left(\frac{\partial \psi}{\partial r}\right)_{x,y} \mathbf{r},$$

where $\mathbf{i}, \mathbf{j}$ are unit vectors parallel to the x, y-axes respectively.

We have

$$\left(\frac{\partial \psi}{\partial x}\right)_{y,z} = \left(\frac{\partial \psi}{\partial x}\right)_{y,r} + \left(\frac{\partial \psi}{\partial r}\right)_{x,y} \frac{\partial r}{\partial x}$$

$$= \left(\frac{\partial \psi}{\partial x}\right)_{y,r} + \frac{x}{r}\left(\frac{\partial \psi}{\partial r}\right)_{x,y}, \tag{1}$$

$$\left(\frac{\partial \psi}{\partial y}\right)_{x,z} = \left(\frac{\partial \psi}{\partial y}\right)_{x,r} + \frac{y}{r}\left(\frac{\partial \psi}{\partial r}\right)_{x,y}, \tag{2}$$

$$\left(\frac{\partial \psi}{\partial z}\right)_{x,y} = \left(\frac{\partial \psi}{\partial r}\right)_{x,y}\left(\frac{\partial r}{\partial z}\right) = \frac{z}{r}\left(\frac{\partial \psi}{\partial r}\right)_{x,y}. \tag{3}$$

Now $\mathbf{r} = x\mathbf{i}+y\mathbf{j}+z\mathbf{k}$ where $\mathbf{k}$ is the unit vector in the z-direction, and we have

$$\begin{aligned}
\operatorname{grad} \psi &= \left(\frac{\partial \psi}{\partial x}\right)_{y,z}\mathbf{i} + \left(\frac{\partial \psi}{\partial y}\right)_{x,z}\mathbf{j} + \left(\frac{\partial \psi}{\partial z}\right)_{x,y}\mathbf{k} \\
&= \left(\frac{\partial \psi}{\partial x}\right)_{y,z}\mathbf{i} + \left(\frac{\partial \psi}{\partial y}\right)_{x,z}\mathbf{j} + \left(\frac{\partial \psi}{\partial z}\right)_{x,y}\left(\frac{1}{z}\mathbf{r} - \frac{x}{z}\mathbf{i} - \frac{y}{z}\mathbf{j}\right) \\
&= \left\{\left(\frac{\partial \psi}{\partial x}\right)_{y,z} - \frac{x}{z}\left(\frac{\partial \psi}{\partial z}\right)_{x,y}\right\}\mathbf{i} + \left\{\left(\frac{\partial \psi}{\partial y}\right)_{x,z} - \frac{y}{z}\left(\frac{\partial \psi}{\partial z}\right)_{x,y}\right\}\mathbf{j} \\
&\qquad + \frac{1}{z}\left(\frac{\partial \psi}{\partial z}\right)_{x,y}\mathbf{r} \\
&= \left(\frac{\partial \psi}{\partial x}\right)_{y,r}\mathbf{i} + \left(\frac{\partial \psi}{\partial y}\right)_{x,r}\mathbf{j} + \frac{1}{r}\left(\frac{\partial \psi}{\partial r}\right)_{x,y}\mathbf{r},
\end{aligned}$$

using (*1*) and (*3*) and (*2*) and (*3*).

16.7. *Calculate the work done by the force field*

$$\mathbf{F} = (2xy^2 + yz)\mathbf{i} + (2yz^2 + zx)\mathbf{j} + (2zy^2 + xy)\mathbf{k}$$

in moving a particle from the point $(0, 0, 0)$ *to the point* $(1, 1, 1)$ *along the curve* $x = t$, $y = t^2$, $z = t^3$. *Show that the field* $\mathbf{G} = \mathbf{F} + 2x^2y\mathbf{j}$ *is irrotational and find its potential function. Calculate the work done by this field in the above displacement.*

$$\begin{aligned}
\text{Work done} &= \int \mathbf{F}\,.\,d\mathbf{r} \\
&= \int_0^1 \{(2t^5 + t^5)(1) + (2t^8 + t^4)(2t) + (2t^7 + t^3)(3t^2)\}\,dt \\
&= \int_0^1 \{8t^5 + 10t^9\}\,dt = \tfrac{7}{3}.
\end{aligned}$$

Now

$$\mathbf{G} = (2xy^2 + yz)\mathbf{i} + (2yz^2 + zx + 2x^2y)\mathbf{j} + (2zy^2 + xy)\mathbf{k},$$

and

$$\text{curl } \mathbf{G} = (4yz + x - 4yz - x, \ y - y, \ z + 4xy - 4xy - z)$$

$$= \mathbf{0},$$

so that the field $\mathbf{G}$ is irrotational.

Let V be the potential function corresponding to $\mathbf{G}$; then

$$\frac{\partial V}{\partial x} = 2xy^2 + yz, \quad \frac{\partial V}{\partial y} = 2yz^2 + zx + 2x^2y, \quad \frac{\partial V}{\partial z} = 2zy^2 + xy.$$

From these relations we get

$$V = x^2y^2 + xyz + \lambda(y, z), \quad V = y^2z^2 + xyz + x^2y^2 + \mu(z, x),$$

$$V = y^2z^2 + xyz + \nu(x, y).$$

Hence $V = y^2z^2 + x^2y^2 + xyz$.

The work done in moving the particle from $(0, 0, 0)$ to $(1, 1, 1)$ under $\mathbf{G}$ is $V(1, 1, 1) - V(0, 0, 0) = 3 - 0 = 3$.

16.8. *Show that*

$$\left| \text{curl } (\phi\mathbf{v}) - \phi \text{ curl } \mathbf{v} \right|^2 + \{\text{div } (\phi\mathbf{v}) - \phi \text{ div } \mathbf{v}\}^2$$

$$= \left| \mathbf{v} \right|^2 \left| \text{grad } \phi \right|^2,$$

where ϕ is a scalar and $\mathbf{v}$ a vector field.

Let θ be the angle at a field point between $\mathbf{v}$ and grad ϕ; then using *(16.13)* and *(16.8)*, the left-hand side of the above identity becomes

$$\left| -\mathbf{v} \times \text{grad } \phi \right|^2 + \{\mathbf{v} \cdot \text{grad } \phi\}^2$$

$$= \left| \mathbf{v} \right|^2 \left| \text{grad } \phi \right|^2 (\sin^2 \theta + \cos^2 \theta) = \left| \mathbf{v} \right|^2 \left| \text{grad } \phi \right|^2.$$

ADDITIONAL EXAMPLES

16.9. Show that, if $\mathbf{a}$ is a constant vector and $\mathbf{r}$ is the position vector (x, y, z) ($\left| \mathbf{r} \right| = r$), then

$$\text{curl } \{r^2 (\mathbf{r} \times \mathbf{a})\} = 2(\mathbf{r} \cdot \mathbf{a})\mathbf{r} - 4r^2\mathbf{a}.$$

16.10. If ϕ is a scalar field satisfying $\nabla^2\phi = 0$ and $\mathbf{r}$ is the position vector show that $\mathbf{u} = \mathbf{r} \times \text{grad } \phi$ satisfies

    (i) curl $\mathbf{u} = -\text{grad }\{\phi+(\mathbf{r}.\text{ grad }\phi)\}$;   (ii) $\nabla^2\mathbf{u} = \mathbf{0}$.

16.11. Establish the identities

    (i) $(\mathbf{u}.\text{ grad})(\mathbf{k}.\mathbf{r}) = (\mathbf{u}.\mathbf{k})$;

    (ii) $(\mathbf{u}.\text{ grad})\mathbf{r} \log r = \mathbf{u} \log r + \dfrac{\mathbf{r}}{r^2}(\mathbf{u}.\mathbf{r})$;

    (iii) $(\mathbf{u}.\text{ grad})\mathbf{u} = \frac{1}{2}\text{ grad }|\mathbf{u}|^2 - \mathbf{u} \times \text{curl }\mathbf{u}$,

where $\mathbf{u}$ is a vector field, $\mathbf{r}$ is the position vector ($|\mathbf{r}| = r$) and $\mathbf{k}$ is a constant vector.

16.12. If $\mathbf{a}$ is a constant vector, $\mathbf{u}$ a vector field and $\mathbf{r}$ is the position vector ($|\mathbf{r}| = r$), show that

    (i) $\mathbf{a} \times \nabla \times (\mathbf{u} \times \mathbf{a}) = \mathbf{a} \times (\mathbf{a}.\nabla)\mathbf{u}$;

    (ii) $\mathbf{a} \times \nabla \times \{\mathbf{a} \times \nabla(1/r)\} = 3(\mathbf{r}.\mathbf{a})(\mathbf{r} \times \mathbf{a})/r^5$.

16.13. The vector fields $\mathbf{E}$ and $\mathbf{H}$ are solenoidal and are related by the equations

$$\text{curl }\mathbf{E} = -\mu\frac{\partial\mathbf{H}}{\partial t}, \quad \text{curl }\mathbf{H} = \varepsilon\frac{\partial\mathbf{E}}{\partial t},$$

where μ, ε are constants. Show that both $\mathbf{E}$ and $\mathbf{H}$ satisfy the equation

$$\nabla^2\mathbf{A} = \varepsilon\mu\frac{\partial^2\mathbf{A}}{\partial t^2}.$$

(A vector field $\mathbf{u}$ is solenoidal when div $\mathbf{u} = 0$.)

16.14. Two vector fields $\mathbf{E}_1$, $\mathbf{E}_2$ are defined by

$$\mathbf{E}_1 = (ye^{xy}+2xyz)\mathbf{i}+(xe^{xy}+x^2z)\mathbf{j}+f(x,y)\mathbf{k},$$

$$\mathbf{E}_2 = (ye^{xy}+2xyz)\mathbf{i}+(xe^{xy}+x^2z)\mathbf{j}+g(y,z)\mathbf{k},$$

where f and g are arbitrary functions. Show that one and only one of the two fields can be made irrotational by means of a suitable choice of the arbitrary function. Find the potential function in this case.

16.15. If $\phi = (\mathbf{a}.\mathbf{r})^n$ where $\mathbf{a}$ is a constant non-zero vector and $\mathbf{r}$ is the position vector, show that ϕ is a solution of Laplace's equation

E

$\nabla^2 f = 0$, if and only if $n = 0, 1$. Show that ϕ is a solution of the biharmonic equation

$$\nabla^4 f = \nabla^2(\nabla^2 f) = 0$$

if and only if $n = 0, 1, 2, 3$.

16.16. If **a** is a constant unit vector, **r** is the position vector $(|\mathbf{r}| = r)$, $f = r(r - \mathbf{a} \cdot \mathbf{r})$ and $\mathbf{w} = \mathbf{r} \times \mathbf{a}$, prove that
(*i*) curl $\mathbf{w} = -2\mathbf{a}$;   (*ii*) grad $f = [2 - r^{-1}(\mathbf{a} \cdot \mathbf{r})]\mathbf{r} - r\mathbf{a}$;
(*iii*) $\mathbf{w} \times \operatorname{grad} f = 2f\mathbf{a} + f^2 r^{-3}\mathbf{r}$;   (*iv*) $\operatorname{curl}\left(\dfrac{\mathbf{w}}{f}\right) = r^{-3}\mathbf{r}$.

16.17. Two vector fields **A** and **B** satisfy the equations

$$\operatorname{curl} \mathbf{A} = c\mathbf{B}, \quad \operatorname{curl} \mathbf{B} = c\mathbf{A},$$

where c is a non-zero constant. Show that

$$\operatorname{div} \mathbf{A} = \operatorname{div} \mathbf{B} = 0,$$

and that **A** and **B** satisfy the equation $\nabla^2 \mathbf{V} + c^2 \mathbf{V} = \mathbf{0}$.
If **X** is any vector satisfying $\nabla^2 \mathbf{X} + c^2 \mathbf{X} = \nabla \phi$ where ϕ is a scalar field, verify that the vectors $\mathbf{A} = c\operatorname{curl} \mathbf{X}$ and $\mathbf{B} = \operatorname{curl} \operatorname{curl} \mathbf{X}$ satisfy the equations

$$\operatorname{curl} \mathbf{A} = c\mathbf{B}, \quad \operatorname{curl} \mathbf{B} = c\mathbf{A}.$$

17. Integral Theorems

In vector calculus it is customary to write the triple integral

$$\iiint_K f(x, y, z)\,dxdydz \quad \text{as} \quad \int_K f(x, y, z)\,d\tau,$$

using only a single integral sign and the volume differential $d\tau$ instead of $dxdydz$. Similarly the surface integral

$$\iint_S f(x, y, z)\,dS \quad \text{is written as} \quad \int_S f(x, y, z)\,dS.$$

In §**14** we discussed Gauss's theorem in three dimensions:

$$\iiint_K \left(\frac{\partial P}{\partial x}+\frac{\partial Q}{\partial y}+\frac{\partial R}{\partial z}\right) dxdydz = \iint (Pdydz+Qdzdx+Rdxdy),$$

where the surface integral is taken over the outside of the surface S. Let (n_1, n_2, n_3) be the direction cosines of the outward normal to S, so that Gauss's theorem can be written as

$$\int_K \left(\frac{\partial P}{\partial x}+\frac{\partial Q}{\partial y}+\frac{\partial R}{\partial z}\right) d\tau = \int_S (Pn_1+Qn_2+Rn_3)dS.$$

Let $\mathbf{n}$ be the vector with components (n_1, n_2, n_3), the unit vector in the direction of the outward normal to S, and let $\mathbf{v}$ be the vector field with components (P, Q, R). With this notation Gauss's theorem becomes

$$\int_K (\text{div } \mathbf{v})d\tau = \int_S (\mathbf{v} \cdot \mathbf{n})dS, \qquad (17.1)$$

and in this form it is usually known as the *divergence theorem*.

Kelvin's (or Stokes's) theorem, as discussed in §15, states

$$\int_C (Fdx+Gdy+Hdz)$$

$$= \iint_S \left\{ l\left(\frac{\partial H}{\partial y}-\frac{\partial G}{\partial z}\right)+m\left(\frac{\partial F}{\partial z}-\frac{\partial H}{\partial x}\right)+n\left(\frac{\partial G}{\partial x}-\frac{\partial F}{\partial y}\right)\right\} dS,$$

where the curve C is the boundary of the surface, (l, m, n) are the direction cosines of the outward normal to a given side of S and the line integral round C is taken in the positive direction associated with this side of S. We shall now replace (l, m, n) by (n_1, n_2, n_3) and let $\mathbf{n}$ be the unit vector (n_1, n_2, n_3) and $\mathbf{v}$ the vector field (F, G, H). We denote the vector (dx, dy, dz) by $d\mathbf{s}$, i.e., a vector in the direction round C indicated above. With this notation Kelvin's theorem becomes

$$\int_C \mathbf{v} \cdot d\mathbf{s} = \int_S (\mathbf{n} \cdot \text{curl } \mathbf{v}) dS. \qquad (17.2)$$

It is sometimes convenient to write $\mathbf{n}dS$ as $d\mathbf{S}$, the vector differential of surface area on the side of S to which $\mathbf{n}$ is the outward normal direction. With this notation (17.1) can be written as

$$\int_K (\text{div } \mathbf{v}) \, d\tau = \int_S \mathbf{v} \cdot d\mathbf{S} \qquad (17.3)$$

and (17.2) as

$$\int_C \mathbf{v} \cdot d\mathbf{s} = \int_S \text{curl } \mathbf{v} \cdot d\mathbf{S}. \qquad (17.4)$$

If in the original Gauss's theorem we put $P = \psi$, $Q = R = 0$, we get

$$\int_K \frac{\partial \psi}{\partial x} \, d\tau = \int_S \psi n_1 \, dS,$$

and we have corresponding results for $\partial \psi / \partial y$ and $\partial \psi / \partial z$. Thus

$$\int_K \left(\frac{\partial \psi}{\partial x}, \frac{\partial \psi}{\partial y}, \frac{\partial \psi}{\partial z} \right) d\tau = \int_S \psi (n_1, n_2, n_3) dS,$$

i.e.,

$$\int_K (\text{grad } \psi) d\tau = \int_S \psi d\mathbf{S}. \qquad (17.5)$$

Again, if we put $P = 0$, $Q = v_3$, $R = -v_2$, we get

$$\int_K \left(\frac{\partial v_3}{\partial y} - \frac{\partial v_2}{\partial z} \right) d\tau = \int_S (n_2 v_3 - n_3 v_2) dS,$$

the first component of the vector relation

$$\int_K (\text{curl } \mathbf{v}) d\tau = -\int_S \mathbf{v} \times d\mathbf{S}. \qquad (17.6)$$

WORKED EXAMPLES

17.1. *Express in terms of the del operator the identities (17.3)–(17.6).*

We use the notation given in (16.7) to get

(17.3)
$$\int_K (\nabla \cdot \mathbf{v})d\tau = \int_S \mathbf{v} \cdot d\mathbf{S}.$$

(17.4)
$$\int_C \mathbf{v} \cdot d\mathbf{s} = \int_S (\nabla \times \mathbf{v}) \cdot d\mathbf{S}.$$

(17.5)
$$\int_K \nabla\psi \, d\tau = \int_S \psi \, d\mathbf{S}.$$

(17.6)
$$\int_K (\nabla \times \mathbf{v})d\tau = -\int_S \mathbf{v} \times d\mathbf{S}.$$

17.2. *Show that, if* **a** *is a constant vector and* **r** *the position vector,*

$$\int_S (\mathbf{r} \times \mathbf{a}) \times d\mathbf{S} = 2V\mathbf{a},$$

where S is the surface bounding a region whose volume is V.

In (17.6) let

$$\mathbf{v} = \mathbf{r} \times \mathbf{a} = (ya_3 - za_2, \ za_1 - xa_3, \ xa_2 - ya_1),$$

where $\mathbf{a} = (a_1, a_2, a_3)$, so that the x-component of curl $(\mathbf{r} \times \mathbf{a})$ is

$$\frac{\partial}{\partial y}(xa_2 - ya_1) - \frac{\partial}{\partial z}(za_1 - xa_3) = -2a_1,$$

and similarly for the other components.
We thus have

$$\int_S (\mathbf{r} \times \mathbf{a}) \times d\mathbf{S} = +2\int_K \mathbf{a} \, d\tau = 2V\mathbf{a}.$$

17.3. *If S is the boundary surface of a region K and if the vector field* **w** *is normal to S at all points on it, show that*

$$\int_K (\text{grad } \phi \cdot \text{curl } \mathbf{w})d\tau = 0,$$

where ϕ is any scalar field.

By (*16.11*)

$$\text{div} (\mathbf{w} \times \text{grad } \phi) = \text{grad } \phi \cdot \text{curl } \mathbf{w} - \mathbf{w} \cdot \text{curl grad } \phi$$

$$= \text{grad } \phi \cdot \text{curl } \mathbf{w},$$

since $\text{curl grad } \phi = \mathbf{0}$, by (*16.9*).   Hence

$$\int_K (\text{grad } \phi \cdot \text{curl } \mathbf{w}) d\tau = \int_K \text{div} (\mathbf{w} \times \text{grad } \phi) d\tau$$

$$= \int_S (\mathbf{w} \times \text{grad } \phi) \cdot \mathbf{n} \, dS, \text{ by } (17.1).$$

Now $\mathbf{w}$ is parallel to $\mathbf{n}$ and thus $\mathbf{w} \times \text{grad } \phi$ is perpendicular to $\mathbf{n}$, so that the surface integral is zero, giving the required result.

17.4. *If* $\mathbf{E}$ *and* $\mathbf{H}$ *are vector fields, varying with time t and related by the equations*

$$\frac{d\mathbf{E}}{dt} = \text{curl } \mathbf{H}, \quad \frac{d\mathbf{H}}{dt} = -\text{curl } \mathbf{E},$$

prove that

$$\tfrac{1}{2} \frac{d}{dt} \int_V (\mathbf{E}^2 + \mathbf{H}^2) d\tau = -\int_S (\mathbf{E} \times \mathbf{H}) \cdot d\mathbf{S},$$

where S is the boundary of the region V.

We have

$$\tfrac{1}{2} \frac{d}{dt} \int_V (\mathbf{E}^2 + \mathbf{H}^2) d\tau = \int_V \left(\mathbf{E} \cdot \frac{d\mathbf{E}}{dt} + \mathbf{H} \cdot \frac{d\mathbf{H}}{dt} \right) d\tau$$

$$= \int_V (\mathbf{E} \cdot \text{curl } \mathbf{H} - \mathbf{H} \cdot \text{curl } \mathbf{E}) d\tau = \int_V \text{div} (\mathbf{H} \times \mathbf{E}) d\tau, \text{ by } (16.11),$$

$$= -\int_S (\mathbf{E} \times \mathbf{H}) \cdot d\mathbf{S}, \quad \text{by } (17.3).$$

17.5. *Show with the usual notation that if ψ is a scalar field*

$$\int_C \psi \, d\mathbf{s} = \int_S d\mathbf{S} \times \operatorname{grad} \psi.$$

Verify the truth of this result when S is the surface

$$x^2 + y^2 + z^2 = a^2 \quad (x \geq 0, \, y \geq 0, \, z \geq 0), \quad \psi = r^2(\mathbf{b} \cdot \mathbf{r}),$$

b *being a constant vector and* **r** *the position vector* $(\,|\,\mathbf{r}\,| = r)$.

Using Kelvin's theorem with $F = \psi$, $G = H = 0$, we have

$$\int_C \psi \, dx = \int_S \left(n_2 \frac{\partial \psi}{\partial z} - n_3 \frac{\partial \psi}{\partial y} \right) dS$$

with similar results for $\displaystyle\int_C \psi \, dy$ and $\displaystyle\int_C \psi \, dz$.

Combining these we get

$$\int_C \psi \, d\mathbf{s} = \int_S d\mathbf{S} \times \operatorname{grad} \psi.$$

In the verification we have for the surface integral

$$(n_1, n_2, n_3) = (x/a, \, y/a, \, z/a) \text{ and}$$

$$dS = \frac{a \, dx \, dy}{\sqrt{(a^2 - x^2 - y^2)}}.$$

Let $\mathbf{b} = (b_1, b_2, b_3)$ so that $\psi = r^2(b_1 x + b_2 y + b_3 z)$.

Then the x-component of $\displaystyle\int_S d\mathbf{S} \times \operatorname{grad} \psi$

$$= \iint \left(n_2 \frac{\partial \psi}{\partial z} - n_3 \frac{\partial \psi}{\partial y} \right) \frac{a \, dx \, dy}{\sqrt{(a^2 - x^2 - y^2)}},$$

over the first quadrant of the circle $x^2 + y^2 = a^2$,

$$= \iint \left\{ \frac{y}{a}(b_3a^2 + 2z(b_1x+b_2y+b_3z)) - \frac{z}{a}(b_2a^2 + 2y(b_1x+b_2y+b_3z)) \right\}$$

$$\times \frac{adxdy}{\sqrt{(a^2-x^2-y^2)}}, \text{ where } z = \sqrt{(a^2-x^2-y^2)},$$

$$= \iint \left(\frac{a^2b_3y}{\sqrt{(a^2-x^2-y^2)}} - b_2a^2 \right) dxdy$$

$$= \iint a^2b_3 \frac{r^2 \sin\theta}{\sqrt{(a^2-r^2)}} \, drd\theta - \frac{\pi}{4}b_2a^4$$

$$= \frac{\pi}{4}a^4(b_3-b_2).$$

The x-component of the line integral is

$$\int_C \psi \, dx = \int_a^0 a^2(b_1x+b_2y)dx + \int_0^a a^2(b_1x+b_3z)dx,$$

where $y = \sqrt{(a^2-x^2)}$ and $z = \sqrt{(a^2-x^2)}$,

$$= \int_0^a a^2(b_3-b_2)\sqrt{(a^2-x^2)}dx = \frac{\pi a^4}{4}(b_3-b_2).$$

We get similar results for the other components and have the required verification.

ADDITIONAL EXAMPLES

17.6. Show, with the usual notation, that

$$\int_S \mathbf{v}(\mathbf{u} \cdot d\mathbf{S}) = \int_K \{\mathbf{v} \, \text{div} \, \mathbf{u} + (\mathbf{u} \cdot \text{grad}) \, \mathbf{v}\} d\tau.$$

17.7. Show, with the usual notation, that

$$\int_K (\mathbf{u} \cdot \text{curl curl } \mathbf{v} - \mathbf{v} \cdot \text{curl curl } \mathbf{u}) d\tau$$

$$= \int_S (\mathbf{v} \times \text{curl } \mathbf{u} - \mathbf{u} \times \text{curl } \mathbf{v}) \cdot d\mathbf{S}.$$

17.8. Show, with the usual notation, that

$$\int_K (\mathbf{r} \,.\, \mathrm{curl}\ \mathbf{u})\,d\tau = \int_S (\mathbf{u} \times \mathbf{r})\,.\,d\mathbf{S}.$$

17.9. Verify that $\displaystyle\int_K (\mathrm{div}\ \mathbf{v})\,d\tau = \int_S (\mathbf{v}\,.\,\mathbf{n})\,dS$ in the case where

$\mathbf{v} = xz\mathbf{i} + zy\mathbf{j}$ and S is the total surface of the right circular cylinder, having height h and radius of base b, whose base is in the xy-plane with axis the z-axis.

17.10. Verify that $\displaystyle\int_S \mathrm{curl}\ \mathbf{v}\,.\,d\mathbf{S} = \int_C \mathbf{v}\,.\,d\mathbf{s}$, where $\mathbf{v} = xy\mathbf{i} + z^3\mathbf{j} + x^3\mathbf{k}$,

and S is the surface given by $x^2 + y^2 = 9$, $0 \leqslant z \leqslant 2$, $x \geqslant 0$, $y \geqslant 0$.

17.11. Show, with the usual notation, that when grad $\phi = \mathrm{curl}\ \mathbf{u}$,

$$\int_S (\mathbf{u} \times \mathrm{curl}\ \mathbf{u})\,.\,d\mathbf{S} = \int_K |\,\mathrm{grad}\ \phi\,|^2\,d\tau.$$

17.12. Show, with the usual notation, that

$$\int_S (\mathbf{u} \times \mathrm{grad}\ \phi)\,.\,d\mathbf{S} = \int_K (\mathrm{grad}\ \phi\,.\,\mathrm{curl}\ \mathbf{u})\,d\tau.$$

Chapter VI

FOURIER SERIES

18. Full-range Fourier Series

Let $f(x)$ be a continuous function or have a finite number of finite discontinuities in the range $(-\pi, \pi)$. Then it can be shown that for $-\pi < x < \pi$,

$$f(x) = a_0 + \sum_{n=1}^{\infty} (a_n \cos nx + b_n \sin nx), \qquad (18.1)$$

where

$$a_0 = \frac{1}{2\pi} \int_{-\pi}^{\pi} f(x)\,dx,$$

$$a_n = \frac{1}{\pi} \int_{-\pi}^{\pi} f(x) \cos nx\,dx, \quad n = 1, 2, 3, \ldots, \qquad (18.2)$$

$$b_n = \frac{1}{\pi} \int_{-\pi}^{\pi} f(x) \sin nx\,dx, \quad n = 1, 2, 3, \ldots .$$

The series on the right of (18.1) is called the *Fourier series* of $f(x)$ in the range $-\pi < x < \pi$. The numbers $a_0, a_1, a_2, \ldots,$ $b_1, b_2, b_3, \ldots,$ are the *Fourier coefficients* for $f(x)$ in the range $-\pi < x < \pi$.

If $f(x)$ has a finite discontinuity at $x = c$, $(-\pi < c < \pi)$, the sum of the series (18.1) is $\frac{1}{2}\{f(c-)+f(c+)\}$. The sum of the series when $x = \pm\pi$ is $\frac{1}{2}\{f(-\pi+)+f(\pi-)\}$.

Since the series on the right is periodic, with period 2π, the coefficients may be determined by expressions of the form (18.2) with the range of integration any interval of length 2π. For example, $f(x)$ may be defined in $(0, 2\pi)$ and the limits of integration in (18.2) become, in this case, $x = 0$ and $x = 2\pi$.

It is sometimes necessary to represent $f(x)$ by a Fourier series throughout a more general interval, say from $x = 0$ to $x = 2l$. We then have

$$f(x) = a_0 + \sum_{n=1}^{\infty} \left(a_n \cos \frac{n\pi x}{l} + b_n \sin \frac{n\pi x}{l} \right), \qquad (18.3)$$

where $a_0 = \dfrac{1}{2l} \displaystyle\int_0^{2l} f(x)\,dx$,

$$a_n = \frac{1}{l} \int_0^{2l} f(x) \cos \frac{n\pi x}{l}\,dx, \quad n = 1, 2, 3, \ldots, \qquad (18.4)$$

$$b_n = \frac{1}{l} \int_0^{2l} f(x) \sin \frac{n\pi x}{l}\,dx, \quad n = 1, 2, 3, \ldots.$$

WORKED EXAMPLES

18.1. *Discuss the case where* $f(x) = \begin{cases} 1, & -\pi < x < 0, \\ x, & 0 < x < \pi. \end{cases}$

We have

$$a_0 = \frac{1}{2\pi} \int_{-\pi}^0 1\,dx + \frac{1}{2\pi} \int_0^\pi x\,dx = \tfrac{1}{2} + \tfrac{1}{4}\pi,$$

$$a_n = \frac{1}{\pi} \int_{-\pi}^0 \cos nx\,dx + \frac{1}{\pi} \int_0^\pi x \cos nx\,dx$$

$$= 0 + \frac{1}{\pi} \left[\frac{1}{n} x \sin nx \right]_0^\pi - \frac{1}{n\pi} \int_0^\pi \sin nx\,dx$$

$$= \frac{1}{n^2\pi}(\cos n\pi - 1) = \frac{1}{\pi} \begin{cases} 0, & n \text{ even}, \\ \dfrac{-2}{n^2}, & n \text{ odd}, \end{cases}$$

$n = 1, 2, 3, \ldots.$

$$b_n = \frac{1}{\pi} \int_{-\pi}^0 \sin nx\,dx + \frac{1}{\pi} \int_0^\pi x \sin nx\,dx$$

$$= \frac{1}{\pi} \left[-\frac{1}{n} \cos nx \right]_{-\pi}^0 - \frac{1}{\pi} \left[\frac{x}{n} \cos nx \right]_0^\pi + \frac{1}{n\pi} \int_0^\pi \cos nx\,dx$$

$$= \frac{1}{\pi n}(-1 + \cos n\pi) - \frac{1}{n} \cos n\pi + 0 = \begin{cases} -\dfrac{1}{n}, & n \text{ even}, \\ \dfrac{1}{n} - \dfrac{2}{\pi n}, & n \text{ odd}, \end{cases}$$

$n = 1, 2, 3, \ldots$ Hence the Fourier series is

$$\frac{1}{2}+\frac{\pi}{4}-\frac{2}{\pi}\left(\frac{1}{1^2}\cos x+\frac{1}{3^2}\cos 3x+ \ldots\right)$$
$$+\left\{\left(1-\frac{2}{\pi}\right)\sin x-\frac{1}{2}\sin 2x+\left(\frac{1}{3}-\frac{2}{3\pi}\right)\sin 3x- \ldots\right\}.$$

Now $\frac{1}{2}\{f(0-)+f(0+)\} = \frac{1}{2}(1+0) = \frac{1}{2}$, so that when we put $x = 0$ in the Fourier series we get

$$\frac{1}{2} = \frac{1}{2}+\frac{\pi}{4}-\frac{2}{\pi}\left(\frac{1}{1^2}+\frac{1}{3^2}+\frac{1}{5^2}+ \ldots\right).$$

Thus $\qquad \dfrac{1}{1^2}+\dfrac{1}{3^2}+\dfrac{1}{5^2}+ \ldots = \dfrac{\pi^2}{8}.$

When $x = \pi$, the sum of the series is

$$\tfrac{1}{2}\{f(\pi-)+f(-\pi+)\} = \tfrac{1}{2}(\pi+1)$$

that is,

$$\frac{(\pi+1)}{2} = \frac{1}{2}+\frac{\pi}{4}+\frac{2}{\pi}\left(\frac{1}{1^2}+\frac{1}{3^2}+\frac{1}{5^2}+ \ldots\right),$$

which agrees with the previously found result, namely, that

$$1/1^2+1/3^2+1/5^2+ \ldots = \pi^2/8.$$

18.2. *Find the Fourier series in the range $0 < x < 2\pi$ for the function $f(x)$ defined by*

$$f(x) = \begin{cases} \sin \tfrac{1}{2}x, & 0 < x < \pi, \\ 0 &, & \pi < x < 2\pi. \end{cases}$$

Sketch the graph of the sum of the series in the range $-3\pi < x < 3\pi$. Deduce from the Fourier series the sums of the series

(i) $\dfrac{1}{3.5}-\dfrac{1}{7.9}+\dfrac{1}{11.13}-\dfrac{1}{15.17}+ \ldots ;$

(ii) $\dfrac{1}{1.3}-\dfrac{3}{5.7}+\dfrac{5}{9.11}-\dfrac{7}{13.15}+ \ldots .$

We have

$$a_0 = \frac{1}{2\pi} \int_0^{2\pi} f(x)\,dx = \frac{1}{2\pi} \int_0^{\pi} \sin \tfrac{1}{2}x\,dx = \frac{1}{\pi},$$

$$a_n = \frac{1}{\pi} \int_0^{2\pi} f(x) \cos nx\,dx = \frac{1}{\pi} \int_0^{\pi} \sin \tfrac{1}{2}x \cos nx\,dx$$

$$= \frac{1}{2\pi} \int_0^{\pi} \{\sin (n+\tfrac{1}{2})x - \sin (n-\tfrac{1}{2})x\}\,dx$$

$$= \frac{1}{\pi} \left(\frac{1}{2n+1} - \frac{1}{2n-1} \right) = -\frac{2}{\pi} \frac{1}{(2n-1)(2n+1)},$$

$$b_n = \frac{1}{\pi} \int_0^{2\pi} f(x) \sin nx\,dx = \frac{1}{\pi} \int_0^{\pi} \sin \tfrac{1}{2}x \sin nx\,dx$$

$$= \frac{1}{2\pi} \int_0^{\pi} \{\cos (n-\tfrac{1}{2})x - \cos (n+\tfrac{1}{2})x\}\,dx$$

$$= \frac{1}{\pi} \left[\frac{1}{2n-1} \sin (n-\tfrac{1}{2})x - \frac{1}{2n+1} \sin (n+\tfrac{1}{2})x \right]_0^{\pi}$$

$$= (-1)^{n-1} \frac{4n}{\pi} \frac{1}{(2n-1)(2n+1)}.$$

Hence the Fourier series is

$$\frac{1}{\pi} + \sum_{n=1}^{\infty} \frac{2}{\pi} \frac{1}{(2n-1)(2n+1)} \{-\cos nx + (-1)^{n-1} 2n \sin nx\}.$$

The graph of the sum of the series in the range $-3\pi < x < 3\pi$ is shown in Fig. 18.

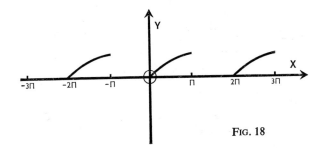

FIG. 18

Now $f(\pi/2) = 1/\sqrt{2}$ and equating this to the sum of the series at this point gives

$$\frac{1}{\sqrt{2}} = \frac{1}{\pi} + \frac{2}{\pi}\left[\left(\frac{1}{3.5} - \frac{1}{7.9} + \frac{1}{11.13} - \cdots\right)\right.$$
$$\left. + 2\left(\frac{1}{1.3} - \frac{3}{5.7} + \frac{5}{9.11} - \cdots\right)\right];$$

when $x = 3\pi/2$ we get

$$0 = \frac{1}{\pi} + \frac{2}{\pi}\left[\left(\frac{1}{3.5} - \frac{1}{7.9} + \frac{1}{11.13} - \cdots\right)\right.$$
$$\left. - 2\left(\frac{1}{1.3} - \frac{3}{5.7} + \frac{5}{9.11} - \cdots\right)\right].$$

Adding gives

(i) $\quad\dfrac{1}{3.5} - \dfrac{1}{7.9} + \dfrac{1}{11.13} - \cdots = \dfrac{\pi}{4\sqrt{2}} - \dfrac{1}{2}$;

subtracting gives

(ii) $\quad\dfrac{1}{1.3} - \dfrac{3}{5.7} + \dfrac{5}{9.11} - \cdots = \dfrac{\pi}{8\sqrt{2}}$.

18.3. *If* $f(x) = \begin{cases} (x-1)^2, & 0 < x < 2, \\ 1, & 2 < x < 4, \end{cases}$ *show that for* $0 < x < 4$

$$f(x) = \frac{2}{3} + \frac{2}{\pi^2}\sum_{n=1}^{\infty}\frac{1}{n^2}\cos n\pi x - \frac{16}{\pi^3}\sum_{n=1}^{\infty}\frac{1}{(2n-1)^3}\sin\frac{(2n-1)\pi x}{2}.$$

With the usual notation, using (18.4), we have

$$a_0 = \tfrac{1}{4}\left\{\int_0^2 (x-1)^2\,dx + \int_2^4 1\,dx\right\} = \tfrac{2}{3},$$

$$a_n = \tfrac{1}{2}\left\{\int_0^2 (x-1)^2\cos\frac{n\pi x}{2}\,dx + \int_2^4 \cos\frac{n\pi x}{2}\,dx\right\}$$

$$= \begin{cases} \dfrac{8}{n^2\pi^2}, & n \text{ even}, \\ 0, & n \text{ odd}, \end{cases}$$

$$b_n = \tfrac{1}{2} \left\{ \int_0^2 (x-1)^2 \sin \frac{n\pi x}{2} \, dx + \int_2^4 \sin \frac{n\pi x}{2} \, dx \right\}$$

$$= \begin{cases} 0 \, , & n \text{ even}, \\ -\dfrac{16}{n^3 \pi^3}, & n \text{ odd}. \end{cases}$$

On writing $2n$ for n in a_n and $2n-1$ for n in b_n, we get the required result.

ADDITIONAL EXAMPLES

18.4. Find the Fourier series, valid for $-\pi < x < \pi$, of the function $f(x)$, where

$$f(x) = \begin{cases} 0, & -\pi < x < 0, \\ x^2, & 0 < x < \pi. \end{cases}$$

18.5. Find the Fourier series, valid for $-\pi < x < \pi$, of the function $x^2 - x$. What are the sums of this series when $x = \pm\pi$? Show that

$$\sum_{n=1}^{\infty} \frac{1}{n^2} = \frac{\pi^2}{6}.$$

18.6. Find the Fourier series, valid for $0 < x < 2\pi$, of the function x^2. Deduce the sum of the series

$$\frac{1}{1^2} - \frac{1}{2^2} + \frac{1}{3^2} - \frac{1}{4^2} + \cdots.$$

18.7. Find the Fourier series, valid for $0 < x < 2$, of the function $f(x)$ where

$$f(x) = \begin{cases} 1, & 0 < x < 1, \\ x, & 1 < x < 2. \end{cases}$$

18.8. If $f(x) = 0$ for $-\pi < x < 0$, $f(x) = 1$ for $0 < x < \pi$, and $f(0) = f(\pm\pi) = 1/2$, show that

$$f(x) = \frac{1}{2} + \frac{2}{\pi} \left\{ \frac{\sin x}{1} + \frac{\sin 3x}{3} + \frac{\sin 5x}{5} + \cdots \right\},$$

and deduce that

(i) $\dfrac{1}{1^2}+\dfrac{1}{3^2}+\dfrac{1}{5^2}+ \ldots = \dfrac{\pi^2}{8},$

(ii) $\dfrac{\cos x}{1^2}+\dfrac{\cos 3x}{3^2}+\dfrac{\cos 5x}{5^2}+ \ldots = \dfrac{\pi(\pi-2x)}{8},$

where $0 \leqslant x \leqslant \pi.$

18.9. Show that, if $-\pi < x < \pi,$

$$1-x \sin x = \tfrac{1}{2}\cos x + \sum_{n=2}^{\infty} (-1)^n \frac{2}{n^2-1} \cos nx.$$

18.10. Show that, if $-\pi < x < \pi,\; a \neq 0,$

$$e^{ax} = \frac{\sinh a\pi}{\pi}\left\{\frac{1}{a}+2\sum_{n=1}^{\infty}(-1)^n\frac{a\cos nx}{a^2+n^2}+2\sum_{n=1}^{\infty}(-1)^{n-1}\frac{n\sin nx}{a^2+n^2}\right\}.$$

By considering the sum of the series when $x = \pi$, deduce that

$$\pi \coth a\pi = \frac{1}{a}+\sum_{n=1}^{\infty}\frac{2a}{a^2+n^2}.$$

SOLUTIONS

18.4. $\dfrac{\pi^2}{6}-2\left(\dfrac{\cos x}{1^2}-\dfrac{\cos 2x}{2^2}+\dfrac{\cos 3x}{3^2}-\dfrac{\cos 4x}{4^2}+ \ldots\right)$

$-\pi\left(\dfrac{\sin 2x}{2}+\dfrac{\sin 4x}{4}+ \ldots\right)$

$+\left(\dfrac{\pi}{1}-\dfrac{4}{\pi 1^3}\right)\sin x+\left(\dfrac{\pi}{3}-\dfrac{4}{\pi 3^3}\right)\sin 3x+\left(\dfrac{\pi}{5}-\dfrac{4}{\pi 5^3}\right)\sin 5x+ \ldots.$

18.5. $\dfrac{\pi^2}{3}+\sum_{n=1}^{\infty}(-1)^n\dfrac{4}{n^2}\cos nx+\sum_{n=1}^{\infty}(-1)^n\dfrac{2}{n}\sin nx;\; \pi^2.$

18.6. $\tfrac{4}{3}\pi^2+\sum_{n=1}^{\infty}\dfrac{4}{n^2}\cos nx-\sum_{n=1}^{\infty}\dfrac{4\pi}{n}\sin nx;\; \dfrac{\pi^2}{12}.$

18.7. $\dfrac{5}{4}+\dfrac{2}{\pi^2}\sum_{n=1}^{\infty}\dfrac{(\cos 2n-1)\,\pi x}{(2n-1)^2}-\dfrac{1}{\pi}\sum_{n=1}^{\infty}\dfrac{\sin n\pi x}{n}.$

19. Half-range Cosine Series

Let $f(x)$ be an even function of x in the range $-\pi < x < \pi$, i.e., $f(-x) = f(x)$. Then the coefficients of the sine terms in the Fourier series are zero and we have the cosine series

$$f(x) = a_0 + \sum_{n=1}^{\infty} a_n \cos nx, \qquad (19.1)$$

where

$$a_0 = \frac{1}{\pi} \int_0^{\pi} f(x)\,dx,$$

$$a_n = \frac{2}{\pi} \int_0^{\pi} f(x) \cos nx\,dx, \quad n = 1, 2, 3, \ldots . \qquad (19.2)$$

Now let $f(x)$ be defined in $0 \leqslant x \leqslant \pi$. Then the series (19.1) with the values of the coefficients given in (19.2) is equal to $f(x)$ in $0 < x < \pi$. This is the *half-range cosine series* for $f(x)$ in $0 < x < \pi$. For the general range $(0, l)$ the half-range cosine series is

$$f(x) = a_0 + \sum_{n=1}^{\infty} a_n \cos \frac{n\pi x}{l},$$

where $\qquad a_0 = \frac{1}{l} \int_0^l f(x)\,dx,$ and

$$a_n = \frac{2}{l} \int_0^l f(x) \cos \frac{n\pi x}{l}\,dx.$$

19.1. *Find the Fourier series representing* $|\sin x|$ *in the range* $(-\pi, \pi)$ *and deduce the sum of the series*

$$\frac{1}{2^2-1} + \frac{1}{6^2-1} + \frac{1}{10^2-1} + \cdots .$$

Since $|\sin x|$ is an even function in $(-\pi, \pi)$ the Fourier series is

$$a_0 + \sum_{n=1}^{\infty} a_n \cos nx, \text{ where}$$

$$a_0 = \frac{1}{\pi} \int_0^\pi \sin x \, dx = \frac{2}{\pi},$$

$$a_n = \frac{2}{\pi} \int_0^\pi \sin x \cos nx \, dx, \quad n = 1, 2, 3, \ldots .$$

Now

$$a_1 = \frac{1}{\pi} \int_0^\pi \sin 2x \, dx = -\tfrac{1}{2}\pi \left[\cos 2x \right]_0^\pi = 0,$$

and if $n = 2, 3, \ldots$

$$a_n = \frac{1}{\pi} \int_0^\pi \{\sin (n+1)x - \sin (n-1)x\} \, dx$$

$$= \frac{1}{\pi} \left[\frac{1}{n-1} \cos (n-1)x - \frac{1}{n+1} \cos (n+1)x \right]_0^\pi$$

$$= \begin{cases} 0 & , \; n \text{ odd,} \\ \dfrac{-4}{\pi(n^2-1)}, & n \text{ even.} \end{cases}$$

Hence, when $-\pi < x < \pi$

$$|\sin x| = \frac{2}{\pi} - \frac{4}{\pi} \sum_{m=1}^\infty \frac{\cos 2mx}{4m^2-1}.$$

When $x = 0$, we have

$$0 = \frac{2}{\pi} - \frac{4}{\pi} \left(\frac{1}{4 \cdot 1^2 - 1} + \frac{1}{4 \cdot 2^2 - 1} + \frac{1}{4 \cdot 3^2 - 1} + \cdots \right)$$

and when $x = \tfrac{1}{2}\pi$, we have

$$1 = \frac{2}{\pi} - \frac{4}{\pi} \left(\frac{-1}{4 \cdot 1^2 - 1} + \frac{1}{4 \cdot 2^2 - 1} + \frac{-1}{4 \cdot 3^2 - 1} + \cdots \right).$$

On subtracting, we get

$$1 = \frac{8}{\pi} \left(\frac{1}{4 \cdot 1^2 - 1} + \frac{1}{4 \cdot 3^2 - 1} + \cdots \right),$$

i.e.,

$$\frac{1}{2^2-1}+\frac{1}{6^2-1}+\frac{1}{10^2-1}+\ldots=\frac{\pi}{8}.$$

19.2. *Find a cosine series for* $f(x) = x$ *in the range* $0 < x < \pi$. *What function does this series represent in the range* $-\pi < x < 0$?

Here $a_0 = \dfrac{1}{\pi}\displaystyle\int_0^\pi x\,dx = \dfrac{\pi}{2},$

$$a_n = \frac{2}{\pi}\int_0^\pi x\cos nx\,dx$$

$$= \frac{2}{\pi}\left[\frac{x}{n}\sin nx\right]_0^\pi - \frac{2}{\pi n}\int_0^\pi \sin nx\,dx$$

$$= \frac{2}{\pi n^2}\left[\cos nx\right]_0^\pi = \begin{cases} 0, & n\text{ even,} \\ \dfrac{-4}{\pi n^2}, & n\text{ odd.} \end{cases}$$

Hence, if $0 < x < \pi$,

$$x = \frac{\pi}{2} - \frac{4}{\pi}\left(\frac{\cos x}{1^2}+\frac{\cos 3x}{3^2}+\frac{\cos 5x}{5^2}+\ldots\right).$$

In the range $-\pi < x < 0$, the series represents the function $f(x) = -x$.

19.3. *Find the Fourier expansion in* $(-\pi, \pi)$ *of the function* $\pi^2 - 3x^2$ *and deduce that of* $x(\pi^2 - x^2)$ *in the same interval. Hence show that*

(i) $\displaystyle\sum_{n=1}^\infty \frac{(-1)^{n-1}}{n^2} = \frac{\pi^2}{12};$ (ii) $\displaystyle\sum_{n=1}^\infty \frac{1}{n^2} = \frac{\pi^2}{6};$

(iii) $\displaystyle\sum_{n=1}^\infty \frac{(-1)^{n-1}}{(2n-1)^3} = \frac{\pi^3}{32}.$

Since $\pi^2 - 3x^2$ is an even function, we have here a cosine series and

$$a_0 = \frac{1}{\pi}\int_0^\pi (\pi^2 - 3x^2)\,dx = \frac{1}{\pi}\left[\pi^2 x - x^3\right]_0^\pi = 0,$$

$$a_n = \frac{2}{\pi} \int_0^\pi (\pi^2 - 3x^2) \cos nx \, dx$$

$$= \frac{2}{\pi} \left\{ \left[\frac{\pi^2}{n} \sin nx \right]_0^\pi - \left[\frac{3x^2 \sin nx}{n} \right]_0^\pi + \int_0^\pi 6x \, \frac{\sin nx}{n} \, dx \right\}$$

$$= \frac{2}{\pi} \left\{ \left[-6x \, \frac{\cos nx}{n^2} \right]_0^\pi + \int_0^\pi 6 \, \frac{\cos nx}{n^2} \, dx \right\}$$

$$= (-1)^{n-1} \frac{12}{n^2}.$$

Hence, when $-\pi < x < \pi$

$$\pi^2 - 3x^2 = 12 \left(\frac{\cos x}{1^2} - \frac{\cos 2x}{2^2} + \frac{\cos 3x}{3^2} - \cdots \right). \tag{1}$$

When $x = 0$, we get at once

$$\text{(i)} \quad \sum_{n=1}^\infty \frac{(-1)^{n-1}}{n^2} = \frac{\pi^2}{12}.$$

Since $f(\pi-) = f(\pi+)$, for the function represented by the series we may put $x = \pi$ in (1) to get

$$-2\pi^2 = 12 \left(\frac{-1}{1^2} - \frac{1}{2^2} - \frac{1}{3^2} - \cdots \right),$$

i.e.

$$\text{(ii)} \quad \sum_{n=1}^\infty \frac{1}{n^2} = \frac{\pi^2}{6}.$$

On integrating both sides of (1) we obtain

$$\int_{-\pi}^x (\pi^2 - 3x^2) dx = \int_{-\pi}^x 12 \left(\frac{\cos x}{1^2} - \frac{\cos 2x}{2^2} + \frac{\cos 3x}{3^2} - \cdots \right) dx,$$

i.e., $\pi^2 x - x^3 = 12 \left(\dfrac{\sin x}{1^3} - \dfrac{\sin 2x}{2^3} + \dfrac{\sin 3x}{3^3} - \cdots \right).$

Hence the series on the right is the Fourier series for $\pi^2 x - x^3$ in $(-\pi, \pi)$. On putting $x = \frac{1}{2}\pi$, we get

$$\frac{3\pi^3}{8} = 12 \left(\frac{1}{1^3} - \frac{1}{3^3} + \frac{1}{5^3} - \cdots \right),$$

i.e. $\qquad$ (iii) $\displaystyle\sum_{n=1}^{\infty} \frac{(-1)^{n-1}}{(2n-1)^3} = \frac{\pi^3}{32}$.

ADDITIONAL EXAMPLES

19.4. Obtain the Fourier series for $\cosh x$ in the interval $-\pi < x < \pi$. Deduce the sums of the series

(i) $\frac{1}{2} - \frac{1}{5} + \frac{1}{10} - \frac{1}{17} + \frac{1}{26} - \cdots$,

(ii) $\frac{1}{5} - \frac{1}{17} + \frac{1}{37} - \frac{1}{65} + \frac{1}{101} - \cdots$.

19.5. Show that the Fourier series for the function x^2 in the range $-\pi < x < \pi$ is

$$\frac{\pi^2}{3} + \sum_{n=1}^{\infty} (-1)^n \frac{4}{n^2} \cos nx.$$

Find the function represented by this series in the range $\pi \leqslant x \leqslant 3\pi$.

19.6. Show that

$$\sin \tfrac{1}{2}x = \frac{2}{\pi} - \frac{4}{\pi} \sum_{n=1}^{\infty} \frac{\cos nx}{4n^2 - 1}, \quad 0 \leqslant x \leqslant \pi.$$

19.7. If

$$f(x) = \begin{cases} 0 \,, & 0 \leqslant x \leqslant \tfrac{1}{2}\pi, \\ x - \tfrac{1}{2}\pi, & \tfrac{1}{2}\pi \leqslant x \leqslant \pi, \end{cases}$$

show that, for $0 \leqslant x \leqslant \pi$,

$$f(x) = \frac{\pi}{8} - \frac{2}{\pi}\left(\frac{\cos x}{1^2} + \frac{\cos 3x}{3^2} + \frac{\cos 5x}{5^2} + \cdots\right)$$
$$+ \frac{4}{\pi}\left(\frac{\cos 2x}{2^2} + \frac{\cos 6x}{6^2} + \frac{\cos 10x}{10^2} + \cdots\right).$$

19.8. If

$$f(x) = \begin{cases} 4x^2, & 0 < x < \tfrac{1}{2}\pi, \\ \pi^2, & \tfrac{1}{2}\pi < x < \pi, \end{cases}$$

show that, for $0 < x < \pi$,

$$f(x) = \frac{2\pi^2}{3} + \sum_{n=1}^{\infty} \frac{8}{n^2} \left(\cos \frac{n\pi}{2} - \frac{2}{n\pi} \sin \frac{n\pi}{2} \right) \cos nx.$$

19.9. If

$$f(x) = \begin{cases} 4 - x^2, & 0 < x < 1, \\ 3, & 1 < x < 2, \end{cases}$$

show that, for $0 < x < 2$,

$$f(x) = \frac{10}{3} + \frac{8}{\pi^2} \sum_{n=1}^{\infty} \left(\frac{2}{n^3\pi} \sin \frac{n\pi}{2} - \frac{1}{n^2} \cos \frac{n\pi}{2} \right) \cos \frac{n\pi x}{2}.$$

19.10. If

$$f(x) = \begin{cases} 1, & 0 < x \leqslant 1, \\ \dfrac{c-x}{c-1}, & 1 \leqslant x < c, \end{cases}$$

show that, for $0 < x < c$,

$$f(x) = \frac{c+1}{2c} + \frac{4c}{\pi^2(c-1)} \sum_{n=1}^{\infty} \frac{1}{n^2} \sin \frac{(c+1)n\pi}{2c} \sin \frac{(c-1)n\pi}{2c} \cos \frac{n\pi x}{c}.$$

SOLUTIONS

19.4. $\dfrac{1}{\pi} \sinh \pi \left\{ 1 + 2 \displaystyle\sum_{n=1}^{\infty} (-1)^n \dfrac{\cos nx}{1+n^2} \right\};$

$(i)\ \dfrac{1}{2} - \dfrac{\pi}{2 \sinh \pi},$ $(ii)\ \dfrac{1}{2} - \dfrac{\pi}{4 \sinh \frac{1}{2}\pi}.$

19.5. $(x - 2\pi)^2$.

20. Half-range Sine Series

Let $f(x)$ be an odd function of x in the range $-\pi < x < \pi$, i.e., $f(-x) = -f(x)$. Then the coefficients of the cosine terms in the Fourier series are zero and we have the sine series

$$f(x) = \sum_{n=1}^{\infty} b_n \sin nx, \qquad\qquad\qquad (20.1)$$

where

$$b_n = \frac{2}{\pi} \int_0^{\pi} f(x) \sin nx \, dx, \quad n = 1, 2, 3, \ldots . \qquad (20.2)$$

Now let $f(x)$ be defined in $0 < x < \pi$. Then the series (20.1) with the values of the coefficients given in (20.2) is equal to $f(x)$ in $0 < x < \pi$. This is the *half-range sine series* for $f(x)$ in $0 < x < \pi$. For general range $(0, l)$ the half-range sine series is

$$\sum_{n=1}^{\infty} b_n \sin \frac{n\pi x}{l}, \quad \text{where} \quad b_n = \tfrac{2}{l} \int_0^l f(x) \sin \frac{n\pi x}{l} \, dx.$$

WORKED EXAMPLES

20.1. *Prove that, for* $-\pi < x < \pi$,

$$\frac{x}{2} = \sin x - \frac{\sin 2x}{2} + \frac{\sin 3x}{3} - \frac{\sin 4x}{4} + \ldots .$$

What is the sum of the series when (i) $x = \tfrac{3}{2}\pi$, (ii) $x = -\tfrac{5}{4}\pi$? *Show that*

$$\frac{1}{1 \cdot 2} + \frac{1}{4 \cdot 5} + \frac{1}{7 \cdot 8} + \ldots = \frac{\pi}{3\sqrt{3}}.$$

We have

$$b_n = \frac{2}{\pi} \int_0^{\pi} \frac{x}{2} \sin nx \, dx = \frac{1}{\pi} \left[-\frac{x}{n} \cos nx \right]_0^{\pi} + \frac{1}{\pi} \int_0^{\pi} \frac{1}{n} \cos nx \, dx$$

$$= (-1)^{n-1} \frac{1}{n}, \quad n = 1, 2, 3, \ldots .$$

Since $\tfrac{1}{2}x$ is an odd function, it follows that for $-\pi < x < \pi$

$$\frac{x}{2} = \sin x - \frac{\sin 2x}{2} + \frac{\sin 3x}{3} - \frac{\sin 4x}{4} + \ldots .$$

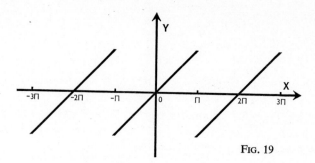

FIG. 19

It is of interest to sketch the graph of the sum of the series for the range $-3\pi < x < 3\pi$, which is shown in Fig. 19.

Hence the sum of the series is (i) $-\dfrac{\pi}{4}$ when $x = \tfrac{3}{2}\pi$ and

(ii) $\tfrac{3}{8}\pi$ when $x = -\tfrac{5}{4}\pi$.

Putting $x = \tfrac{1}{3}\pi$ in the series, we get

i.e.,

$$\frac{\pi}{6} = \frac{\sqrt{3}}{2} - \frac{\sqrt{3}}{2 \cdot 2} + \frac{\sqrt{3}}{2 \cdot 4} - \frac{\sqrt{3}}{2 \cdot 5} + \frac{\sqrt{3}}{2 \cdot 7} - \frac{\sqrt{3}}{2 \cdot 8} + \dots,$$

$$\frac{\pi}{3\sqrt{3}} = \left(\frac{1}{1} - \frac{1}{2}\right) + \left(\frac{1}{4} - \frac{1}{5}\right) + \left(\frac{1}{7} - \frac{1}{8}\right) + \dots$$

$$= \frac{1}{1 \cdot 2} + \frac{1}{4 \cdot 5} + \frac{1}{7 \cdot 8} + \dots .$$

20.2. *Show that, for $0 < x < \pi$,*

$$\cos x = \frac{4}{\pi}\left\{\frac{2\sin 2x}{1 \cdot 3} + \frac{4\sin 4x}{3 \cdot 5} + \frac{6\sin 6x}{5 \cdot 7} + \dots\right\}.$$

What is the sum of this series for $-\pi < x < 0$?

Here

$$b_n = \frac{2}{\pi}\int_0^\pi \cos x \sin nx \, dx,$$

so that,

$$b_1 = \frac{1}{\pi}\int_0^\pi \sin 2x \, dx = \tfrac{1}{2}\pi\left[-\cos 2x\right]_0^\pi = 0,$$

and for $n = 2, 3, \ldots,$

$$b_n = \frac{1}{\pi} \int_0^\pi \{\sin(n+1)x + \sin(n-1)x\} dx$$

$$= \frac{1}{\pi} \left[-\frac{\cos(n+1)x}{n+1} - \frac{\cos(n-1)x}{n-1} \right]_0^\pi$$

$$= \begin{cases} 0, & n \text{ odd,} \\ \dfrac{4n}{(n-1)(n+1)\pi}, & n \text{ even.} \end{cases}$$

Hence for $0 < x < \pi$

$$\cos x = \frac{4}{\pi} \left\{ \frac{2\sin 2x}{1 \cdot 3} + \frac{4\sin 4x}{3 \cdot 5} + \frac{6\sin 6x}{5 \cdot 7} + \ldots \right\}.$$

For $-\pi < x < 0$, the sum of the series is $-\cos x$.

20.3. *Find the half-range Fourier sine series of the function* $x(\pi - x)$ *in the interval* $(0, \pi)$. *What is the sum of the series in the interval* $(\pi, 2\pi)$?

Show that

$$(i) \quad \frac{1}{1^3} - \frac{1}{3^3} + \frac{1}{5^3} - \frac{1}{7^3} + \ldots = \frac{\pi^3}{32},$$

$$(ii) \quad \frac{1}{1^3} + \frac{1}{3^3} - \frac{1}{5^3} - \frac{1}{7^3} + \ldots = \frac{3\sqrt{2}\pi^3}{128}.$$

Here

$$b_n = \frac{2}{\pi} \int_0^\pi (\pi x - x^2) \sin nx \, dx$$

$$= \left[\frac{2}{n\pi} (x^2 - \pi x) \cos nx \right]_0^\pi + \frac{2}{n\pi} \int_0^\pi (\pi - 2x) \cos nx \, dx$$

$$= \left[\frac{2}{n^2\pi} (\pi - 2x) \sin nx \right]_0^\pi + \frac{4}{n^2\pi} \int_0^\pi \sin nx \, dx$$

$$= \frac{4}{n^3\pi} \left[-\cos nx \right]_0^\pi = \begin{cases} 0, & n \text{ even,} \\ \dfrac{8}{n^3\pi}, & n \text{ odd.} \end{cases}$$

Hence for $0 < x < \pi$

$$x(\pi - x) = \frac{8}{\pi}\left[\frac{\sin x}{1^3} + \frac{\sin 3x}{3^3} + \frac{\sin 5x}{5^3} + \cdots\right].$$

The sum of the series in $(\pi, 2\pi)$ is $(x-\pi)(x-2\pi)$. When $x = \frac{1}{2}\pi$, we have

$$\frac{\pi}{2}\left(\frac{\pi}{2}\right) = \frac{8}{\pi}\left(\frac{1}{1^3} - \frac{1}{3^3} + \frac{1}{5^3} - \frac{1}{7^3} + \cdots\right).$$

i.e. $(i)\ \dfrac{1}{1^3} - \dfrac{1}{3^3} + \dfrac{1}{5^3} - \dfrac{1}{7^3} + \cdots = \dfrac{\pi^3}{32};$

when $x = \frac{1}{4}\pi$, we have

$$\frac{\pi}{4}\left(\frac{3\pi}{4}\right) = \frac{8}{\pi}\left[\frac{1}{\sqrt{2}.1^3} + \frac{1}{\sqrt{2}.3^3} - \frac{1}{\sqrt{2}.5^3} - \frac{1}{\sqrt{2}.7^3} + \cdots\right],$$

i.e., $(ii)\ \dfrac{1}{1^3} + \dfrac{1}{3^3} - \dfrac{1}{5^3} - \dfrac{1}{7^3} + \cdots = \dfrac{3\sqrt{2}\pi^3}{128}.$

20.4. *If*

$$f(x) = \begin{cases} 2x + x^2, & -2 < x \leqslant 0, \\ 2x - x^2, & 0 \leqslant x < 2, \end{cases}$$

prove that, for $-2 < x < 2$,

$$f(x) = \frac{32}{\pi^3} \sum_{n=1}^{\infty} \frac{1}{(2n-1)^3} \sin(2n-1)\frac{\pi x}{2}.$$

Since $f(x)$ is an odd function, the Fourier series is a sine series and we can deduce it from the last example, where, if

$$\phi(x) = \begin{cases} x(\pi + x), & -\pi < x \leqslant 0, \\ x(\pi - x), & 0 \leqslant x < \pi, \end{cases}$$

$$\phi(x) = \frac{8}{\pi} \sum_{n=1}^{\infty} \frac{1}{(2n-1)^3} \sin(2n-1)x.$$

Let $x = \frac{1}{2}\pi z$ and we get that the function defined by $\frac{1}{2}\pi z(\pi + \frac{1}{2}\pi z)$ for $-2 < z \leqslant 0$ and $\frac{1}{2}\pi z(\pi - \frac{1}{2}\pi z)$ for $0 \leqslant z < 2$ has Fourier series

$$\frac{8}{\pi} \sum_{n=1}^{\infty} \frac{1}{(2n-1)^3} \sin (2n-1) \frac{\pi z}{2},$$

and hence the function defined by $z(2+z)$ for $-2 < z \leqslant 0$, $z(2-z)$ for $0 \leqslant z < 2$ has series

$$\frac{32}{\pi^3} \sum_{n=1}^{\infty} \frac{1}{(2n-1)^3} \sin (2n-1) \frac{\pi z}{2}.$$

On writing x for z, we get the required result.

ADDITIONAL EXAMPLES

20.5. Show that, if $-\pi < x < \pi$,

$$x \cos x = -\tfrac{1}{2} \sin x + \sum_{n=2}^{\infty} (-1)^n \frac{2n}{n^2-1} \sin nx.$$

20.6. Show that, if $-\pi < x < \pi$,

$$\sin \tfrac{1}{2}x = \frac{8}{\pi} \sum_{n=1}^{\infty} (-1)^{n-1} \frac{n}{4n^2-1} \sin nx.$$

20.7. Show that, if $-\pi < x < \pi$,

$$\frac{\pi}{2} \frac{\sinh ax}{\sinh a\pi} = \sum_{n=1}^{\infty} (-1)^{n-1} \frac{n \sin nx}{n^2+a^2}.$$

Deduce that, if $-\pi \leqslant x \leqslant \pi$,

(i) $\dfrac{\pi}{2a} \dfrac{\cosh ax - 1}{\sinh a\pi} = \displaystyle\sum_{n=1}^{\infty} (-1)^{n-1} \dfrac{1 - \cos nx}{n^2+a^2},$

(ii) $\dfrac{\pi}{\sinh a\pi} = \dfrac{1}{a} + \displaystyle\sum_{n=1}^{\infty} (-1)^n \dfrac{2a}{n^2+a^2}.$

20.8. If

$$f(x) = \begin{cases} p-x, & 0 \leqslant x \leqslant 2p, \\ x-3p, & 2p \leqslant x \leqslant 4p, \end{cases}$$

find a half-range sine series for $f(x)$ in the range $0 < x < 4p$. What is the sum of the series when $x = 6p$?

20.9. If

$$f(x) = \begin{cases} 1, & 0 < x < \pi, \\ -1, & -\pi < x < 0, \\ 0, & x = 0, \pi, -\pi, \end{cases}$$

express $f(x)$ as a Fourier series in the interval $(-\pi, \pi)$ and show that

$$\sum_{n=1}^{\infty} \frac{\sin(2n-1)x}{2n-1} = \frac{\pi}{4}, \quad 0 < x < \pi.$$

Deduce that

$$\sum_{n=1}^{\infty} \frac{\cos(2n-1)x}{(2n-1)^2} = \frac{\pi}{4}\left(\frac{\pi}{2}-x\right), \quad 0 \leqslant x \leqslant \pi.$$

What is the value of the series on the left of the last relation when (i) $-\pi \leqslant x \leqslant 0$; (ii) $\pi \leqslant x \leqslant 2\pi$? Find the sum of the series

$$\frac{1}{1^3} - \frac{1}{3^3} + \frac{1}{5^3} - \frac{1}{7^3} + \cdots.$$

20.10. If

$$f(x) = \begin{cases} \pi+x, & -\pi < x < 0, \\ \frac{1}{2}\pi, & x = -\pi, 0, \pi, \\ x, & 0 < x < \pi, \end{cases}$$

show that, for $-\pi \leqslant x \leqslant \pi$

$$f(x) = \frac{1}{2}\pi - \sum_{n=1}^{\infty} \frac{\sin 2nx}{n}.$$

Deduce that, if $0 \leqslant x \leqslant \pi$,

$$\sum_{n=1}^{\infty} \frac{\cos 2nx}{n^2} = x^2 - \pi x + S,$$

where

$$S = \sum_{n=1}^{\infty} \frac{1}{n^2}.$$

By putting $x = \frac{1}{2}\pi$, prove that $S = \frac{1}{6}\pi^2$ and deduce that

(i) $\displaystyle\sum_{n=1}^{\infty} \frac{\sin 2nx}{n^3} = \frac{1}{3}x(\pi-x)(\pi-2x), \quad 0 \leqslant x \leqslant \pi;$

(ii) $\dfrac{1}{1^3} - \dfrac{1}{3^3} + \dfrac{1}{5^3} - \dfrac{1}{7^3} + \ldots = \dfrac{\pi^3}{32}.$

SOLUTIONS

20.8. $\dfrac{4p}{\pi}\left(\dfrac{1}{1} - \dfrac{4}{1^2\pi}\right)\sin\dfrac{\pi x}{4p} + \dfrac{4p}{\pi}\left(\dfrac{1}{3} + \dfrac{4}{3^2\pi}\right)\sin\dfrac{3\pi x}{4p}$

$$+\dfrac{4p}{\pi}\left(\dfrac{1}{5} - \dfrac{4}{5^2\pi}\right)\sin\dfrac{5\pi x}{4p} + \ldots ; \; p.$$

20.9. (i) $\dfrac{\pi}{4}\left(x + \dfrac{\pi}{2}\right).$ (ii) $\dfrac{\pi}{4}\left(x - \dfrac{3\pi}{2}\right); \dfrac{\pi^3}{32}.$

21. Parseval's Theorem

If $\{a_n, b_n\}$ are the Fourier coefficients of a function $f(x)$ defined in $(-\pi, \pi)$, Parseval's theorem states that

$$\int_{-\pi}^{\pi} [f(x)]^2 \, dx = 2\pi a_0^2 + \pi \sum_{n=1}^{\infty} (a_n^2 + b_n^2).$$

The theorem can be used to evaluate certain infinite series.

WORKED EXAMPLES

21.1. *Use the Fourier series for* $|\sin x|$ *in the range* $(-\pi, \pi)$ *to find the sum of the series.*

$$\dfrac{1}{(2^2 - 1)^2} + \dfrac{1}{(4^2 - 1)^2} + \dfrac{1}{(6^2 - 1)^2} + \ldots .$$

In **19.1** we saw that in $(-\pi, \pi)$,

$$|\sin x| = \dfrac{2}{\pi} - \dfrac{4}{\pi}\sum_{m=1}^{\infty} \dfrac{\cos 2mx}{4m^2 - 1}$$

so that $a_0 = 2/\pi$, $a_n = -4/\pi(n^2 - 1)$ (n even), $a_n = 0$ (n odd).

Now

$$\int_{-\pi}^{\pi} [f(x)]^2 \, dx = 2 \int_{0}^{\pi} \sin^2 x \, dx = \pi.$$

Hence

$$\pi = 2\pi \left(\frac{2}{\pi}\right)^2 + \pi \left(\frac{16}{\pi^2(2^2-1)^2} + \frac{16}{\pi^2(4^2-1)^2} + \cdots\right)$$

and so

$$\frac{1}{(2^2-1)^2} + \frac{1}{(4^2-1)^2} + \frac{1}{(6^2-1)^2} + \cdots = \frac{1}{16}(\pi^2-8).$$

21.2. *Use the Fourier series for the function* $\pi^2 - 3x^2$ *in the range* $(-\pi, \pi)$ *to show that*

(i) $\displaystyle\sum_{n=1}^{\infty} \frac{1}{n^4} = \frac{\pi^4}{90}$;   (ii) $\displaystyle\sum_{n=1}^{\infty} \frac{1}{n^6} = \frac{\pi^6}{945}$.

In **19.3** we saw that in $(-\pi, \pi)$

$$\pi^2 - 3x^2 = 12 \left(\frac{\cos x}{1^2} - \frac{\cos 2x}{2^2} + \frac{\cos 3x}{3^2} - \cdots\right).$$

Now

$$\int_{-\pi}^{\pi} (\pi^2 - 3x^2)^2 dx = 2 \int_{0}^{\pi} (\pi^4 - 6\pi^2 x^2 + 9x^4) dx = \tfrac{8}{5}\pi^5.$$

Hence

$$\frac{8\pi^5}{5} = \pi \left(\frac{144}{1^4} + \frac{144}{2^4} + \frac{144}{3^4} + \cdots\right)$$

and so

$$\sum_{n=1}^{\infty} \frac{1}{n^4} = \frac{\pi^4}{90}.$$

In **19.3** we deduced from the series for $\pi^2 - 3x^2$ that

$$\pi^2 x - x^3 = 12 \left(\frac{\sin x}{1^3} - \frac{\sin 2x}{2^3} + \frac{\sin 3x}{3^3} - \cdots\right)$$

in $(-\pi, \pi)$. Hence, since

$$\int_{-\pi}^{\pi} (\pi^2 x - x^3)^2 dx = 2 \int_{0}^{\pi} (\pi^4 x^2 - 2\pi^2 x^4 + x^6) dx = \tfrac{16}{105}\pi^7,$$

we have

$$\frac{16\pi^7}{105} = 144\pi \left(\frac{1}{1^6} + \frac{1}{2^6} + \frac{1}{3^6} + \cdots \right),$$

i.e.,

$$\sum_{n=1}^{\infty} \frac{1}{n^6} = \frac{\pi^6}{945}.$$

21.3. *Use the half-range sine series for $x(\pi-x)$ in the range $(0, \pi)$ to find the sum of the series*

$$\frac{1}{1^6} + \frac{1}{3^6} + \frac{1}{5^6} + \frac{1}{7^6} + \cdots .$$

In **20.3** we saw that

$$x(\pi-x) = \frac{8}{\pi} \left(\frac{\sin x}{1^3} + \frac{\sin 3x}{3^3} + \frac{\sin 5x}{5^3} + \cdots \right).$$

Now

$$2 \int_0^\pi x^2(\pi-x)^2 dx = 2 \int_0^\pi (\pi^2 x^2 - 2\pi x^3 + x^4)dx = \tfrac{1}{15}\pi^5.$$

Hence

$$\frac{\pi^5}{15} = \frac{64}{\pi} \left(\frac{1}{1^6} + \frac{1}{3^6} + \frac{1}{5^6} + \cdots \right),$$

i.e.,

$$\frac{1}{1^6} + \frac{1}{3^6} + \frac{1}{5^6} + \cdots = \frac{\pi^6}{960}.$$

ADDITIONAL EXAMPLE

21.4. Use the Fourier series in **20.2** to find the sum of the series

$$\left(\frac{2}{1 \cdot 3} \right)^2 + \left(\frac{4}{3 \cdot 5} \right)^2 + \left(\frac{6}{5 \cdot 7} \right)^2 + \cdots .$$

SOLUTION

21.4. $\tfrac{1}{16}\pi^2$.